T0073441

PHYSICS AND OUR WORLD

Reissue of the Proceedings of a Symposium
in Honor of Victor F Weisskopf

PHYSICS AND OUR WORLD

Reissue of the Proceedings of a Symposium
in Honor of Victor F Weisskopf

Editor

Kerson Huang

Massachusetts Institute of Technology, USA

 **World Scientific**

NEW JERSEY · LONDON · SINGAPORE · BEIJING · SHANGHAI · HONG KONG · TAIPEI · CHENNAI

Published by

World Scientific Publishing Co. Pte. Ltd.

5 Toh Tuck Link, Singapore 596224

USA office: 27 Warren Street, Suite 401-402, Hackensack, NJ 07601

UK office: 57 Shelton Street, Covent Garden, London WC2H 9HE

British Library Cataloguing-in-Publication Data
A catalogue record for this book is available from the British Library.

The editor and publisher would like to thank the American Institute of Physics for permission to reproduce the articles, figures, and the photograph found in this book from:
AIP Conference Proceedings, No. 28
"Physics and Our World: A Symposium in Honor of Victor F. Weisskopf"
© 1976 American Institute of Physics

PHYSICS AND OUR WORLD
Reissue of the Proceedings of a Symposium in Honor of Victor F Weisskopf

ISBN 978-981-4434-96-6

Typeset by Stallion Press
Email: enquiries@stallionpress.com

Printed in Singapore by World Scientific Printers.

Contents

Foreword

This is a reissue of *Physics and Our World: A Symposium in Honor of Victor F. Weisskopf, MIT, 17–18 October, 1974,* ed. K. Huang (American Institute of Physics, New York, 1976), originally published as *AIP Conference Proceedings No. 28.* We thank the American Institute of Physics for permission to reset and reprint this volume. The figures, the speakers' original sketches, have been preserved.

Victor Weisskopf (1908–2002), professor of physics at MIT, known to his friends as "Viki", was an influential figure in physics, both for his work as a theoretical physicist, and science statesman. The prominence of the symposium speakers, and the thoughtfulness of their speeches, testify to the high regard and great affection in which he was held in the physics community.

Many contributions are a delight to read. Hans Bethe's analysis of the energy problem could be written today, but for the giveaway: "Last year, the cost of a barrel of oil in the Middle East was about $3. Now, it is $10 or more." And: "With great difficulty one can find out that one barrel of oil is 42 gallons." Today, of course, oil is around $100 a barrel, and you can instantly find out how many gallons are in a barrel from Google.

Edward Purcell's piece on how one should swim in a viscous fluid — by turning anything, as long as it isn't perfectly symmetrical — tells bacteria to develop flagella.

Stanley Ulam pointed out that mathematicians deduce theorems from axioms, but physicists go the other way — try to find the axioms (natural laws) that lead to given theorems (observations). He also recounted his experience with Enrico Fermi on computer simulations of anharmonic oscillators (which led to an astounding non-ergodic result later attributed to solitons), and Fermi was quoted as saying, "Well, it does not say in the Bible that the fundamental equations of physics must be linear."

Max Delbruck said that Aristotle's physics was "pretty much a catastrophe", but his biological thinking pointed to DNA, for he had the idea that what one generation passes to the next is a blueprint. "Aristotle considered it remarkable that human beings beget human beings, and not rabbits or an ear of corn."

The historical significance of this Symposium is that it occurred during a paradigm change in theoretical physics, because of quarks. Here, Julian Schwinger's talk "Model-free views of deep inelastic scattering" echoed the "old school", while Murray Gell-Mann's "The world as quarks, leptons, and bosons" was precursor to the new.

After the great breakthrough that was renormalized QED in 1947, there followed a long period of disenchantment with quantum field theory, because the expected quick conquest of the strong interactions did not occur. On the one hand, Lev Landau declared that the renormalized charge of the electron is zero according to quantum field theory (a problem known as "triviality"), which should therefore be buried, "with honors". On the other hand, Geoffrey Chew and his school hold that the basic entities of quantum field theory, elementary particles, do not exist; the hadrons are "bound states of one another". In this view of "nuclear democracy", these particles are resonances in some kind of primordial goo that "bootstraps" itself into existence. Even Gell-Mann, who coined the name "quark", had to tiptoe around with the so-called "current algebra", a somewhat messy scheme which might be characterized as "quarks without quarks".

The first break came with the deep inelastic scattering experiment at SLAC, carried out by Taylor, Friedman, and Kendall in 1968. Richard Feynman interpreted the result in terms of "partons" — point-like "parts" of the nucleon that produce back-scattering, much like Rutherford's atomic nuclei. The clincher came in 1974, just after this Symposium, with the discovery of the J/Psi particle by the group of Samuel Ting at the Brookhaven National Laboratory, and that of Burton Richter at SLAC. This extremely long-lived particle is a non-relativistic bound state of the heavy charmed quark and its antiquark — a veritable "hydrogen atom" whose energy levels unmistakably identify its underlying structure. This left no doubt that quarks do exist, and soon led to the standard model, a renormalizable quantum field theory involving Yang–Mills gauge fields with spontaneous symmetry breaking.

As a historical note, Samuel Ting was present at this Symposium, and his group had already found an extremely narrow resonance peak. In Ting's words, "It was like stumbling into a village in which the average age of the inhabitants was ten thousand years." He had considered announcing the result at the Symposium, but did not, out of carefulness.

With the standard model, quantum field theory again reigns supreme. The bootstrap theory strikes back, however. It morphs into string theory, and challenges quantum field theory when the distance scale shrinks to the Planck scale, by some eighteen orders of magnitude.

Kerson Huang
Naples, Florida
June, 2013

Acknowledgments

We are happy to acknowledge the encouragement and help which we received from President Jerome B. Wiesner, Provost Walter A. Rosenblith, and Head of the Physics Department, Professor Herman Feshbach, of the Massachusetts Institute of Technology.

We also wish to thank all of the speakers at this Symposium for their graciousness in granting permission to publish their talks, and in many cases, reading and editing transcriptions of their taped presentation.

Thanks are due to Mr. Barry Friedman and Mr. Patrick Hayes for editorial help.

Special thanks are due to Mr. Daniel H. Gould, Executive Officer of the M.I.T. Physics Department, for his help and cooperation.

Last but not least, thanks must go to Miss Diane Eulian for her conscientious, excellent, and indispensable effort in the preparation of the manuscript.

The Editor

Preface

The essays contained in this volume are a partial record of the celebration in honor of Victor Weisskopf held at the Massachusetts Institute of Technology, October 17 and 18, 1974. No attempt was made to organize these presentations about a particular area of physics but rather the speakers were asked to select a topic which, in their opinion, would be of great interest to Viki.

The illustrious list of contributors is an indication of the high regard in which he is held by the scientific community. This is in part a consequence of his impact on the development of modern physics through many seminal contributions: the theory of line breadth; the quantizations of the scalar field; quantum electrodynamics, including a first calculation of the Lamb shift; nuclear reactions, including the optical model; the foundations of the shell model; and the structure of hadrons. It is also a consequence of his actions as a statesman of science in his roles as Director-General of CERN, as head of the M.I.T. physics department, as chairman of the High Energy Physics Advisory Committee (HEPAP) and as spokesman for basic science in the public arena, of his efforts on behalf of international cooperation and of his deep appreciation of the human aspects and human qualities of scientific research and education. But of equal importance is his style and spirit so beautifully captured by his own phrase "the joy of insight". There is an unquenchable desire to understand the essential physical elements involved in phenomena, to strip away the complexities of a detailed explanation and to make visible the underlying ideas and concepts. It is an attitude which we at M.I.T. particularly have come to appreciate, and emulate.

He more than satisfied Milton's definition of a great man: "He alone is worthy of the appellation who either does great things or teaches how they may be done, or describes them with a suitable majesty when they have been done."

Herman Feshbach
October, 1975

1

Introductory Remarks

James Killian
Massachusetts Institute of Technology

Honored guest, ladies and gentlemen. We at M.I.T. rejoice that so many of you from near and far are with us these two days in honoring our cherished colleague, Victor Weisskopf. I say Victor but that does not carry the appropriate ring — the appropriate ring of affection. I am reminded of a story about the late Charles Evans Hughes, once the Chief Justice of the U.S. Supreme Court and candidate for President, a man of formidable accomplishments in his profession and in public life. One day his young granddaughter boldly asked him what she should call him, and the Chief Justice replied: "Those who love me call me Charlie". And so, despite his formidable array of titles, responsibilities, accomplishments, honors, and accolades, we all feel more comfortable in addressing Victor as "Viki".

In the Gilbert and Sullivan operetta *Iolenthe*, there is a line which says: "All questions of party are merged in a frenzy of love and devotion," and so it is today for Viki. I am not one competent to speak of Viki's scientific accomplishments — they are amply known and appreciated by this gallery. Here at MIT he is Mr. Science. One of his colleagues has spoken of him as possessing the Copenhagen spirit, the Niels Bohr-like capacity so thoroughly to understand a complex scientific problem, that he could achieve an illuminating simplicity in presenting it or discussing it with all of us. Without yielding any of the rigor and imagination of his scientific work, he has also given meaning both to the scientific and humanistic values of science for his student and colleagues. By his largeness and liberality of spirit he has drawn about him a constellation of scholars both scientist and non-scientist who have found in his values and his clarity of mind sustanance and reassurance in these days when science has come under attack from so many different directions. In the spirit of Copenhagen he has spoken eloquently to the layman about the role of science and, to use the title of one of his books, about knowledge and wonder and the natural world as man knows it.

In 1963 at the centennial of the National Academy of Science, Robert Oppenheimer spoke of the need to help non-scientists understand science. He said: "We the scientists have a modest part to play in history, and the barriers between us and the man of affairs — the statesman, the artist, the lawyer with whom we should be talking — could perhaps be markedly reduced if some of them knew a little more of what we were up to, knew it with some pleasure and some confidence." Despite Viki's heavy burdens of teaching, researching and administering, he is doing precisely this in many different ways. His concern about the role of science and society leads me today to give a brief report on some recent efforts in the United States to regain for science its proper place in our national policymaking.

As you know, the last administration dismantled the President's Science Advisory Committee (some of the former members of that Committee are here today), and the Office of Science and Technology. As a substitute, he has asked the Director of the National Science Foundation to serve as Science Advisor to the President, reporting not directly to the President but to an intermediary. These arrangements, in the judgement of a Committee of the National Academy of Science that was appointed to examine into the possibilities of improving scientific advice to the President, are not adequate to deal with the problems that fact the people who have to make top policy in our Government today. We feel, in this Committee of which I have the honor to be Chairman, that we need some new arrangement — not the old arrangement we had, not a repetition of PSAC, but some new arrangement that can make available adequate scientific and technological advice to the President and his immediate associates in the Executive Office of the President. The advice called for is not merely the exercise of judgement in the White House, the exercise of judgement of choice among well-defined choices of alternatives. Rather the best scientific and technological talent must be immediately involved in identifying new alternatives for the President. These judgements in turn require an understanding of prospective developments at the frontier of science and technology. This is even more true now than it was after Sputnik. I think the need is much greater than it was in 1957 when Eisenhower established PSAC as his immediate scientific advisory committee. The fundamental thesis of the NAS committee, in its report, is that the process of summation and policy-making that takes place at the level of the Presidency requires the accessibility of scientific, techno-logical and engineering counsel — immediate accessibility. Some people in Government today say this is no longer necessary because science is present in all of the departments and agencies of the Government, but this is not an adequate substitute for having available at the top policy level adequate scientific counsel and advice and analysis. There should be means included in the staff structure of the Executive Office of the President to provide a source of scientific and technological analysis and judgement directly to the President, but also to the immediate agencies that he turns to within his family of advisors in the White House.

The first recommendation of the Committee, therefore, was that a coun-cil for science and technology be created within the Executive Office of the President and that this Council be composed of at least three full-time members drawn from the science, engineering and related fields. Now, this was not the first time that a recommendation for such a council has been made. I think Lee Dubridge was the first person that I know who proposed that there be the creation of a Council somewhat analogous to the Council of Economic Advisors. These three or more scientists and engineers would be appointed by the President with the advice and consent of the Senate and would serve at the pleasure of the President. One member would be designated by the President to be Chairman and would bear the responsibil-ity to report to the President. In effect, he would be the President's Science Advisor.

Our Committee proposed that this Council be created by legislation, although it made this quite tentative, because I think it is very important, if such an agency is created, that the President feel it is his own mechanism to serve his own needs. The Committee further recommended that the Coun-cil for Science and Technology be empowered and able to draw from the best talent within the nation's scientific and technological and engineering communities, both from within and without the Government. It should be able to appoint panels of specialists, and to draw heavily on departments and agencies of Government themselves, as well as on the resources of industry, the universities, non-profit centers, the National Academy of Sciences and of Engineering, and the Institute of Medicine. The Committee felt that this Council should have strong working relationships (and this we felt to be very important) with the agencies and other offices of the Executive Office, notably the Domestic Council and the National Security Council. The Chairman of the Council, we recommended, should be a member of the domestic council and should have an active role in the work of the National Security Council as PSAC did at one time. It is particularly important as we look at the role of the domestic council that it has better scientific inputs than it is getting at the present time. The National Security Council must organize its work in a fashion that will best serve the President in accord with his preferred manner of dealing with National Security Affairs.

Consistent with that requirement, there should be a provision for systematically introducing into the work of the NSC the judgement of a scientist and technologist. The Committee hoped that the Council, if proposed, would be looked to, as was PSAC, for provisions of the nation's best scientific and technical knowledge and judgement — particularly in those matters that involve advanced technology of the insights and early warnings — that the scientists working at the frontiers of their specialties would be qualified to transmit. In this fashion the NSC can benefit from highly professional judgements on military technology and arms control as undistorted by jurisdictional lines of thought as they can be. I can't help but recall that it was by virtue of the fact that the President's Science Advisor under Eisenhower was sitting in on a meeting of the National Security Council, when discussions came up about the problem of fallout; and it was possible to report on that occasion that PSAC had reached the conclusion that progress had been made in one routine test, and that it might be desirable to re-examine this. As a result of this, a committee was promptly appointed by the President, of which Hans Bethe was chairman, that made a report to the National Security Council. This started down the long road that ultimately led to the treaty of limitation of nuclear tests. That sort of thing is exceedingly important today, no less than it was at that time. I also think it important to note that, at the present time, the President does not have available to him critical analysis of military proposals that come up from the Department of Defense, and that this is something that is badly needed and that I suspect that the President is going to find he badly needs it.

In a speech to the United Nations on April 14 of this year, Henry Kissinger said, "In a global economy of physical scarcity, science and technology are becoming our most precious resource; no human activity is less national in character than the field of science; no development effort offers more hope than joint scientific and technical cooperation." That was a statement from the Secretary of State. Within this context, Dr. Kissinger directed his remarks to the needs of the developing countries. Relationship with the developed countries as well are affected deeply by the developments of science and technology as they work upon the international scene, and America's position on that scene. International relationships and their aspects are involved in bilateral and multilateral agreements involving science and technology. This Council that has been proposed, working with the National Security Council and the Secretary of State, could help generate and respond to presidential initiatives, to attack mutual problems, through international scientific and technological cooperation. Incidentally, at the present time, unless I am not up-to-date, there is no scientific advisory mechanism available in the Department of State. Again the old PSAC essentially performed that role. It is also true that the Office of Management and Budget should have a strong scientific input in its deliberations and dealings with the budget and while it has greatly strengthened its staff to deal with scientific matters it still very much needs the objective advice that a group could bring to it, if it were properly related to a council of the kind that this Committee proposes. The NAS Committee also proposed that the Council for Science and Technology present to the Congress an annual report on major developments in science and engineering of significance for national policy. This report would be presented first to the President and through him to the Congress. The Committee also offered a corollary suggestion that there should be a capability, within the executive office of the President, for long range policy research and analysis, which would examine continuously the longer-range implications of current budget decisions and other policies, and would seek to anticipate problems that will face the President and the Congress in future years. The Executive Office of the President needs to employ many of the new techniques that have been developed through scholarly work for policy research and analysis. It needs to understand the close relationship for these new techniques through the methods and spirit of the physical, biological and behavioral sciences. And our Committee believed that a way can be found for making these techniques continuously useful to the President. This report, of which I have given a brief summary, is now under consideration by the Executive Office of the President and by appropriate committees in the Congress. Hearings have been held in both the Senate and the House. There are some of you here who testified at those hearings. There seems to be a cordial reception to the proposal. There has been an extraordinary response across the country in the press, and there has been dozens of editorials, well-reported stories, about the recommendations of the report. There seems to be a national understanding of the need.

I summarize the NAS Committee report by way of pointing to the need today for better ways for the scientific and engineering communities of the nation to make their talents and insights available. Both of the plans which I have reported on, involving both scientific advice and analysis, focus on service to the Government, rather than arrangements to seek more support for the Government for science and technology, however important that may be. If there can be adequate policy imports and adequate analysis made available to the Government, I think the chances are that science itself will be well-served, and that the relationship between Government and science will be on a sounder basis, with the result that science itself will benefit in the long run. Just by way of comment indicating the interest that this problem is currently evoking, let me read the lead paragraph from a recent column by Scottie Reston in the New York Times:
(this was on October 10)

If ever there was a time when the President of the United States needed the help of the best objective scientific minds to help him grapple with the problems of food, fuel, transportation, housing and many other things, it is now. But he is a little shorthanded at the present time.

2

The Energy Problem

Hans Bethe
Cornell University, Ithaca, New York

When I was preparing my lecture, I thought there was something wrong with it, namely, that it had very little to do with Viki. However, Dr. Killian, fortunately, gave a lecture with which mine has a lot to do, namely, it points out the great need for science in our present energy situation.

Last year, the cost of a barrel of oil in the Mid-East was about $3.00. Now, it is $10.00 or more. Now, that may seem rather trivial. Very few of us want to buy a barrel of oil, but it is not at all trivial for the balance of payments of nations, of all western nations, of all the industrialized nations, and even more, of the underdeveloped nations. The World Bank has estimated that we shall have to pay — we meaning the oil importing nations — have to pay something like a hundred billion dollars a year to the oil exporting countries. This amount of money is enormous if it is in international trade. It would not be very much inside the United States. But to see its importance in international trade, just compare it with the gold reserve of this country which is something over ten billion dollars. There will, un-doubtedly, be a great shift of wealth from the countries which are now rich to the oil exporting countries. Now, such shifts of riches have taken place before. There was a shift from Spain to England in the sixteenth and seventeenth century, from England to America in the first half of the twentieth century, from America to Japan and Germany in the last decade or two. But the shift that is impending is of a very different nature. In all previous shifts of wealth, the country that was now getting rich wanted something from the country that was previously rich. They wanted services and goods. We are now faced with the problem that the countries which will acquire this enormous amount of money do not really have any demand for goods which is commensurate with the amount of riches they will acquire.

What will the oil exporting countries do with their wealth? The best, for all purposes, will be if they use their money to acquire industries and perhaps real estate in the western countries and, in particular, in the United States. While this may lead to foreign domination of our industry in a rather short time, a much worse possibility is that they may leave their money as a short-time deposit in international banks, and in national banks, shifting it from one country to another, mostly to get maximum interest, but in some cases, to cause mischief. It is very much to be hoped that the oil exporting countries see their great interest in preserving the structure of the industrial world which is really the structure in which they themselves enjoy their wealth and in which they themselves can prosper. But, there is no guarantee at all that this will be so. So far, the oil exporting countries have played their advantage to the hilt. There are some slight signs of a change. It may very well be that this tremendous indebtedness, amounting to a trillion dollars ten years from now, of the western countries to the oil exporting countries may lead to a complete breakdown of the industrial and economic system of the west. In my opinion, there has only been one situation in my lifetime which was equally serious as the present one and that was the Second World War. In the Second World War, scientists, and particularly the scientists at this institution, M.I.T., came to the rescue and helped the nation and the western world to survive the crisis. This we need again.

We must find other sources of energy other than oil and natural gas. Natural gas, as you know, is already in very scarce supply and has been for some time. Not only for ourselves must we do this, but even more for Europe, Japan and still more for the underdeveloped nations. Now, to look at the dimensions of the problem, look at Table 1. This table gives information I found most difficult to obtain and most important, namely, a comparison between the various units that people use. People, unfortunately, use a Btu which doesn't mean anything to me. But, fortunately, it is very close to a kilojoule which does mean something to me. A million Btu makes about a hundred kilowatt hours of electricity at a normal efficiency of a generating plant. In global projections, people like to use the Q as a unit

Table 1.

1 Btu	= 1.05 kj	
1 MBtu	= 117 kwh (40% eff.)	
1 quad	= 10^{15} Btu	
1 barrel oil	= 42 gallons	= 150 kg
1 barrel oil	= 6 MBtu	
1 Mbbl/day	= 2.10 mQ/yr.	
1 k cu. ft. nat. gas	= 1 MBtu	
1 ton coal	= 24 MBtu	

Table 2. 1973 net oil exports and imports (MB/d). Also balance of payment at $10/barrel (billions of $).

	Net export	Income		Net import	Cost
Mid-East	20.0	73	West. Europe	14.0	51
Africa	4.8	17	U.S.	6.3	23
Carribean	3.2	12	Japan	5.2	19
		102			93

of energy, or the milli-Q or quad which is 10^{15} (a quadrillion) Btu and that's the unit I'm going to use. With great difficulty one can find out that one barrel of oil is 42 gallons. It is in no dictionary and it corresponds to about six million Btu. One million barrels per day, we use about 17, corresponds to about two quad per year. One thousand cubic feet of natural gas is about one million Btu. To confuse the issue further, people who sell natural gas use the letter "M" for a thousand which we use for a million. One ton of coal is about 24 million Btu.

Table 2 lists the sources of oil.

The production of oil in the United States in 1972 was 26 quad, and the petroleum council projects that with great effort and with using all possible new sources we might get to 28 in 1985. The imports were 12 in 1972; they are rather more now. In 1985, a certain paper wanted to reduce this to eight. I would hope we can reduce it to zero and leave the oil which is available on the world market to those countries which cannot produce any oil themselves, namely, Western Europe, Japan and the underdeveloped countries. Arab exports were 32q. This might be raised by about 25 percent according to statements of the Arab countries. The world consumption was about 87; if it continues the way it has been going, it will be 175 in 1985. A barrel of oil was $4.00 and the present cost is $10.00. The cost of American imports will increase even if we decrease the imports, and would increase to sixty billion dollars if we were to continue to cover all the deficit in energy by oil which is clearly intolerable. The price of world consumption will go up from sixty to about three hundred billion dollars.

Table 3* gives the U.S. consumption of various sources of energy. The total in 1972 was 72 quads, and in 1985 I am sure it will be more than 85 quads. Of this, there was only one quad nuclear, two hydro which is a source which cannot be increased much, so the rest was all fossil fuel of which the majority was made up by oil.

Gas, I think, is a wasting asset. The second column of the Table "1985 High Estimate", is from a Cornell report which was put together last fall. This projection assumed that the oil consumption would stay about the same. The gas consumption, they assumed, might increase more, that is, we might discover more gas. I consider this very doubtful. The rest would have to be covered by coal. I made a lower estimate for 1985 in which I assumed that energy consumption will increase only 2% per year. The high estimate assumes the same increase which has prevailed, 4%. My 2% includes 1% increase in population so there is only 1% per year increase in per capita consumption. Whether we can keep down to such a small increase is very doubtful to me. In this case, I also assume that nuclear power will contribute less than the Cornell report assumed optimistically, simply because the plants will be built more slowly. And still, we need a coal production which is more than twice what we had in 1972.

Table 4 gives the available resources in the world, now in Q, i.e. in thousands of quads. Two geologists are quoted here, McKelvey who is an

Table 4.   Available resources (thousands of Q).

	McKelvey	Hubbert	U.S.
Coal	320	190	23.0
Oil	23	15	1.5
Gas	20	10	1.0
Shale oil	80	10	6.0
Total	450	220	30.0

---

*Table 3 does not appear in this text.

optimist and Hubbert who is a pessimist. The difference in numbers is a factor of two, and one might hope that perhaps the world will settle down at a population of ten billion people — now it is four — and that they will each consume about as much energy as we consume now. In that case, the yearly consumption will be three Q, and that means that on the optimist's assumption all the available fossil fuel will be good for a little over one hundred years. On the pessimist's assumption, for seventy years.

Figure 1 gives the distribution of sources of energy and of uses of energy. This comes from a book by Hottel and Howard here at M.I.T.[†] which is a very important book. It shows that coal, in 1970, gave us only 20% of our energy, oil gave nearly half, gas most of the rest. Most of the oil goes for transportation, as you can imagine. Most of the coal goes for electricity generation. Much of the gas goes for heating of houses and of so-called commercial buildings. Transportation is the one area in which

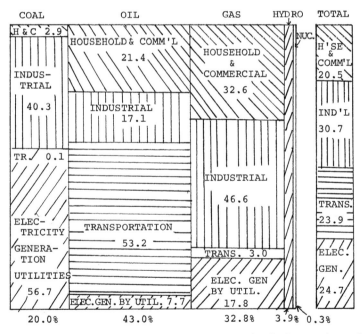

Figure 1.   Distribution of U.S. energy consumption, 1970 (Preliminary estimates by U.S. Department of the Interior).

[†]From New Energy Technology.

oil is practically indispensable. So I think the endeavor must be to shift all the other uses, industrial and especially generation of electricity to other sources of energy.

The report of the Cornell workshop concluded that it is possible to double or even triple coal production in ten or fifteen years. This will be a very major effort. It will obviously mean mostly strip mining, and we have to see to it that after strip mining the land is restored to real fertility and to its previous state, not necessarily in contour but at least in fertility. This is certainly possible and the cost of it is a fraction of the value of the coal that can be obtained from western mines.

Now what do you do with the coal? One possibility is conversion into other substances which are easier to use and to transport. One of them is gas and the other oil. It seems that this is possible. The Germans did it during the Second World War but the cost of their process is high. Table 5 gives the cost of making a million Btu of energy. By the German process it is about $2.00 and with learning it might be reduced a little bit. The present cost of a million Btu of gas, delivered to one of the transcontinental gas pipelines, is $.43. From this you can see that it would pay us very handsomely to permit higher cost of the gas delivered to the pipelines if this gives an incentive to the producers to look for a better solution, maybe

Table 5.

Natural fuels

Cost per MBtu (dollars) — Dec. 1974

	1972	1974
Coal	0.20	0.40–0.80
Oil, Import	0.50	1.80
Oil, Texas	0.65	1.20
Gas	0.20	0.50–1.50

Synthetic fuels

	1985 (1974 dollars)
Oil shale	1.30
Oil from coal	2.50
High Btu gas	2.00
Low Btu gas	1.20
Electric heating	3.50

not to make coal into gas but to transport the energy as electricity. This requires work on better ways to transmit electric power. There is one way I know about, namely super-conducting DC transmission lines. These have been worked on at the Los Alamos Scientific Laboratory for several years. I think they have great promise for the future, and perhaps the most costly thing in this connection is the conversion of AC to DC and vice versa. This problem, as well as the transmission lines, I present as a challenge to this engineering institution. Now, as I showed you, even coal will not last forever. And coal is, in addition to being a source of energy, an increasingly important source of materials. If we run out of oil, we need the coal to make plastics and all sorts of other things which are now called petrochemicals. Therefore, we'd better save some for that purpose and don't use it all up for energy.

The other main source for energy we have is nuclear energy. At present it is a small contribution. In time, I hope, it will be a great deal more. There are clearly three advantages of nuclear energy apart from its availability. The first is, you have minimum environmental trouble. In contrast to coal, the mining of uranium involves small amounts of material and strip mining will not make a major upheaval in the environment. Secondly, nuclear energy is essentially free of pollution in contrast to coal. There are ways to make coal freer of pollution than it is now, but it would be very difficult to make the pollution as small as it would be from a nuclear plant. The third is that even at present the price of nuclear energy is lower than any other energy, and this in spite of the fact that the cost of building a nuclear plant has risen from about $100 per kilowatt installed to about $600.

Now there are many objections to nuclear energy and you hear far more about the objections than about the advantages. One of the objections is that nuclear plants may emit radioactive rays. This, I think, has been solved. The prescription of the Atomic Energy Commission today is that the amount of radiation at the fence posts of the nuclear plant must be less than five millirem per year. Never mind what a millirem is. You are exposed in your ordinary life to about a hundred millirem per year, part of which comes from cosmic rays, part from the radioactivity in the walls of every building, and a large part from yourself, from the potassium that is inside you. Compared to this hundred, the five millirem is a very small amount, and this only occurs at the fence posts — if you live ten miles away from the plant, there will be nothing. The second objection is the high price, which I have already mentioned.

The third is the possibility of an uncontainable accident. Nobody can deny that such a possibility exists. There has been a very intensive study going on over two years led by Professor Rasmussen of this institution which has gone into the probability and the effects of a nuclear accident more carefully than any other study that has previously been made. They come out with the result that the probability is extremely low. If we have a hundred reactors going, the probability of a major release of radioactivity is less than one in a thousand years. Also, if there were an uncontainable accident, they have investigated the possible effects and come to the conclusion that since such an accident is a rather slow affair — release of radioactivity takes a considerable time — it is likely that the number of people who will be in danger, the number of deaths, will be relatively small, in most accidents, something like two, and even in the very rare, major accident, of the order of a hundred. The material damage may be very considerable and in this very rare major accident, it may be as much as several billion dollars, clearly an amount that calls for insurance from the Government rather than for payment from the utility involved. But money is recoverable. They compare the probability of nuclear accidents with that of other catastrophes, natural and manmade, and nuclear accidents come out very well.

Now the opponents of nuclear power must have their day in court and will undoubtedly attack this report. However, if the report is anywhere near right and the care with which it was made, I think, encourages me, then we do not have to fear very much.

The next point is waste disposal. The AEC is now prescribing that all the radioactive waste has to be made into solid material, glassy substances, in contrast to liquid storage, which in the past has given rise to some troubles. Even if the solid were to be destroyed, it is most unlikely that it would get dispersed in a way to make it dangerous. The solid would be stored in strong containers, in an engineered facility, far away from human habitation, let us say in the Nevada desert, and must, of course, be well-guarded so that no fool goes there and gets himself exposed. I don't think that the waste disposal is really a major problem.

The one major problem which I see is the possibility of diversion, the possibility of theft of nuclear materials and use of these materials for bombs, either by some countries or by terrorists or criminals of other kinds. This,

I think, is the most serious problem. This is a problem to which we have to apply all our ingenuity. As far as terrorists and individuals are concerned, I think it can be solved. As far as countries are concerned, I think we have to have administrative ways to minimize the problem.

Clearly, there are problems with fission energy, so let us have a look at other possible sources of energy. The one that is most talked about is fusion. Fusion research started about in 1951 with the highest hopes. Such high hopes that Admiral Strauss who was soon afterwards Chairman of the Atomic Energy Commission admonished scientists working on fusion to have it all finished by the end of the first Eisenhower administration in 1956. You all know that this didn't happen. I won three bets on this. The problem turned out to be far more difficult than anybody envisaged. The plasma which people wanted to use to confine deuterium and tritium to produce fusion simply wouldn't stay confined. It had instabilities far beyond anything that anybody had dreamed of.

What are the conditions necessary for fusion? Well, first you have to have something like one hundred million degrees temperature equivalent to ten thousand electron volts; second you have to have a sufficiently high density; and third the plasma has to stay together a sufficiently long time that the product of the particle density and the time of confinement is at least 10^{14} (e.g., 10^{14} particles in a cubic centimeter has to stay together for one second). At that point, you can get net energy gain from a deuterium–tritium mixture. When theorists saw the instabilities, they first investigated the problem theoretically and found, yes, indeed, these instabilities should be there. Then they sat down a few more years and said well, there are some ways to cure the instabilities. They found one way, and on this basis Artsimovitch in Russia invented a device known as TOKAMAK which uses the theoretical method to reduce instabilities.

Indeed, this device worked to the amazement of everybody who had worked in the field. It is now being used in this country and it still works. The TOKAMAK is a device for magnetic confinement of high temperature, moderately high density plasma so that deuterium and tritium can react with each other. The process looks better than it ever has before. It has been possible to compress that plasma by the magnetic field, and it has been possible to heat it by injection of much higher energy, two to five hundred keV natural deuterium atoms. It has been possible to get fair agreement between theoretical confinement times and experimental ones.

The disagreement is only a factor of ten now, it used to be a factor of a million or more. Partly this is due to the theorists becoming more modest, namely, they have discovered that there are certain instabilities which are unavoidable. Now, the theorists predict that if you make the device much bigger, like ten meters instead of one meter, then it will work. Maybe they are right. Anyway, the AEC has made the money available for constructing a much bigger device, and maybe in 1982 it will be possible to prove that fusion is scientifically feasible. At that time, fusion will be in the state that fission was in 1942 when Fermi proved the first chain reaction. Today, we are not yet in 1942 as far as fusion is concerned. It took twenty years, beginning in 1942, before fission energy became industrially useful.

In fusion energy, we know a great deal beforehand, but it is a much more difficult problem. Radiation damage is relatively minor in the case of fission because you mostly deal with slow neutrons which cause very little damage. In the breeder it is somewhat more. In fusion it will be enormously more because the neutrons have an energy of 14 MeV when they come out. Damage is done by high energy neutrons which propel atoms from any material to God knows where and thereby disintegrate the material which is used for the structure of the device. Furthermore, any fusion device which uses magnetic confinement uses magnetic coils which have to be especially carefully protected. This means you have to locate them far away from the fusion; therefore the magnetic field will extend over an enormous volume which is very expensive. So, I think if we estimate the time from feasibility to industrial usefulness as twenty years this is very modest; from there it will be another twenty years before fusion contributes an appreciable amount to the energy supply of this and other countries. In other words, you get to the year 2020. So it would be entirely irresponsible to rely on fusion as satisfying our energy needs for the twentieth century.

On the other hand, I think research on fusion has to be pursued with great vigor. It may be an important way of getting energy. It would give us an inexhaustible source of energy from the deuterium in the sea and would get us away from *some* of the problems I mentioned. Don't believe, however, that you will get away from the problems of storing radioactive wastes. Wherever there are neutrons there is radioactivity. The radioactivity from fusion may be smaller than it is from fission, and we may be able, to some extent, to choose the radioactive species that we deal with, but there will still be radioactivity also from fusion.

Another kind of fusion has been more fashionable in recent times, namely laser fusion. Lasers are very sexy, fusion is very sexy and so the combination is most sexy. However, laser fusion, I think, in time scale is related to magnetic confinement fusion almost the same way as that is related to fission. It's a very young subject and, while some successes have been achieved, it is not likely to lead to complete success in any short time. What is needed in this case is to take a small pellet of solid deuterium–tritium (that's not very hard to get, you only need a low temperature of some ten degrees Kelvin) and you drop that little pellet into a place where it is exposed to maybe twelve laser beams of very high intensity coming at it from all directions. And then, these laser beams are supposed to compress this pellet to a density between three and ten thousand times its normal density. This I think will be very difficult to realize, the problem is to get enough symmetry, enough freedom from instabilities, enough push in a short enough time. We don't have the lasers to do it. We need much higher power lasers, releasing their energy in a much shorter time, and we need powerful lasers somewhere in the visible which we do not now have at all. Once you have succeeded, your little pellet will give you an energy of about one kilowatt-hour which has a value of about two cents. I do not believe this is a practical device for the foreseeable future. Fifteen years from now we may talk differently, but it will not be the solution of our short-range energy problems.

Well, then there is solar power. Solar power is everywhere and it is a very big power source. In fact, it is many, many times the power which the world may consume in the twenty-first century. However, it is very diffuse and you have to catch it, and catch it reliably. So, in the first place, you have to go to a state where the sun shines most of the time, such as Arizona, maybe New Mexico, maybe western Texas. You have to go to a place where the difference between winter and summer is not too great, so you have to go to the southern states. This gets you back to the problem of transmission of electric power over large distances, but that we have to solve anyway. In addition you have the problem of storage of energy which is much more difficult, from day to night, from sunny to cloudy days which occur even in New Mexico. This is a major problem in which technical development is extremely important.

How can we catch the solar power? From a square kilometer of area, on an average sunny winter day in a southern state, you may get something like 200 megawatts solar heat, which will give perhaps 50 megawatts electric power.

Many fancy ways have been suggested to utilize the solar power. The only way which I think has a good chance of working is a brute force, elementary way, as follows: You build a field of mirrors, each mirror perhaps a square meter, with the field maybe two kilometers on the side. There is a computer drive on the mirrors which directs the sun at any one moment to a central tower, maybe a hundred meters high, on which you have an ordinary boiler, which drives a conventional electric generator. This scheme I think has worked in Southern France. The computer drive should not be very difficult to make. The mirrors are arranged in such a way that they don't shade each other. From your four square kilometers you might then get something like 200 megawatts electric power, average. One of the troubles is that at noon you need an electric plant which is able to generate 800, so that you have 200 in the average. This is one of the reasons why this device will be very expansive. The number that has been mentioned by a fairly responsible engineering source is $2,500 per installed kilowatt, which is about five times the cost of nuclear power at present.

Now, you say what's money? Money means labor and materials, so we cannot afford these large amounts of money even if money seems to be unimportant. For instance, the estimate is that by the year 2000 we need an installed power of a thousand gigawatts, a trillion watts. This would cost over two trillion dollars to install. Therefore, I do not believe that solar power at the cost that we can now foresee is a real solution.

Some people, even the President, have mentioned the winds. I think the winds are for the birds. There is not enough power in the winds. Then there is geothermal power. This, I think, is for real. You can tap hot water reservoirs in the West and possibly even hot rocks, so I think research on geothermal is useful and may even lead to some contribution in the twentieth century.

Having now excluded most of the alternatives, I come back to nuclear power. In my opinion, we have no choice whatever. We must use nuclear fission power; we have to make it as safe as we can, which includes safety against diversion.

Now we come to the question, do we have enough uranium? Table 6 gives the requirements of uranium for several kinds of reactors. The present reactors are light water reactors, first column, and the high temperature gas cooled reactors, second column. Future reactors are the liquid metal

Table 6. Short-term U_3O_8 required to inventory and operate 1,000-MWe reactors for 40 years.

	Inventory (tons)	40-year operation (tons)
PWR	548	5,000
BWR	580	5,000
HTGR	456	2,400
HWR	350	700

fast breeder reactor which requires more inventory for each reactor, and the Canadian-sponsored heavy water reactor which has the nice name "Can–du". Deuterium–Uranium is the meaning of the last two letters. The consumption per year in short tons of uranium ore is very high for the light water reactor, considerably better for the high temperature gas cooled. The fast breeder makes additional fuel and the Candu consumes very little. If you use the light water reactor, exclusively, and that is what we now have, we need about seven and a half million tons of uranium ore by the year 2020 according to projections. If we use a minimum of light water reactors as has been here assumed, namely, until 1990, and afterwards use either the fast breeder or the Candu then we may get away with from three to four million tons of uranium ore.

Figure 2 gives the projected need for power and the projection of the reactors to satisfy it. This Figure I made in connection with a study last year in which the extra need was to be satisfied by fast breeders.

Estimated future nuclear generating capacity, in *millions* of megawatts (electric). AEC figures until 2000; after this, the total electric generating capacity is assumed to double every 25 years. All new capacity is assumed to be nuclear after 2010. Contribution of various types of reactors is shown separately. For detailed assumptions, see Table G2.

Table 7 is the counterpart to Table 6; it gives uranium reserves as presently projected. At present, we use very rich uranium ore which costs only about $10.00 a pound. Of this, there are available about one million tons. Remember, seven million tons is needed if we continue to use light water reactors. Going down to $30.00 uranium doesn't really make the cost of nuclear power much higher but makes the environmental cost considerably higher because you have to dig about eight times more rock than you have to do at the present time. And that about doubles the amount of

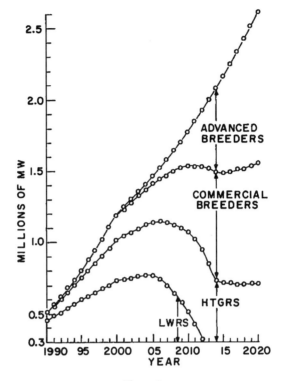

Figure 2.

Table 7.   United States uranium reserves.[a]

Ore concentration ppm U_3O_8	$/lb U_3O_8	1,000 tons U_3O_8		
		Reasonably assured	Estimated additional	Total
1600	Up to $10[b]	427	700	1,127
1000	Up to $15[b]	630	1,000	1,630
200	Up to $30[c]	800	1,600	2,400
60	Up to $50[c]	4,800	3,600	17,400

[a] *Source*: WASH-1242, 1243.
[b] Includes copper leach residues and phosphates.
[c] Includes Chattanooga shale.

available uranium. You have to get down to the uranium listed at $50 in the Table, but this should really be priced at about $100.00 because of the need for restoring the environment. At this level, we have enough uranium ore in this country to satisfy the needs for light water reactors up to the year 2020 and, of course, we have plenty if we go to the more uranium conserving reactors like the Candu or the fast breeder.

I conclude from this that not only do we have to have nuclear energy but we have to find nuclear reactors which are more conservative in the use of uranium. The one that is most talked about is, of course, the breeder, cooled by liquid metal. This has been under development for a long time. An existence proof has been given by the French who have built a fast breeder which works smoothly and reliably and isn't even terribly expensive. Our fast breeder, unfortunately, gets more and more expensive and has now arrived at a cost equal to even solar power, something like $3,000 per kilowatt for the first fast breeder. But this will come down with time, while in the case of solar power no similar decrease of price is in sight.

The reactor which I like best is the heavy water reactor which has been developed by the Canadians. It has been developed with the view of using natural uranium and avoiding the isotope separation. This, I believe, is not the important point for the future. Isotope separation will probably get cheaper than it is now, for instance, by the centrifuge method of separation. But it is important that Candu reactor in Toronto has been available more than ninety percent of the time for electricity generation which is very much better than the average power plant whether nuclear or fossil fired. The other important point is that Candu can be modified to conserve uranium. The modification consists of putting in thorium as a fertile material from which can be bred uranium-233. This advanced Candu can easily be arranged in such a way that you only use ten percent of the fuel you put in. Thus it has a conversion ratio of 90 percent, and, thereby, you get the good performance that showed in Table 7. With ten percent fuel used, our uranium will last a long time. And once uranium ore gets more expensive, Candu will be as economical as our present light water reactors. It costs a little more to build because the heavy water is an expensive substance, but this will easily be repaid in the uranium economy. It is believed, in addition, that it is even safer than our present light water reactors.

The Canadian project, I think, is inviting the United States to participate in the further development of the Thorium thermal breeder with heavy water moderation. I believe that this is the best source of energy that we can think of for the next 50 years.

3

Model-Free Views of Deep Inelastic Scattering

Julian Schwinger
U.C.L.A.

Perhaps I should point out first that my choice of topic was dictated by the injunction that the nature of this symposium should revolve around subjects that might be conceivably of interest to Viki. Viki has, along with most high energy physicists been very interested in the subject of deep inelastic electron scattering. With his characteristic attention to directly visualizable approaches to physical phenomena, he has dealt with this in terms of rather specific models, attempting then to give very elementary explanations of these fascinating phenomena. I thought he might be interested to see the other side of the coin, namely, the extent to which one can correlate and comprehend these physical effects without the use of specific models. I think this may lend a certain useful balance to the way things are looked at these days. So my remarks are directed to Viki but you're all welcome to eavesdrop.

You must forgive me for this somewhat technical subject, and, in fact, I was rather disconcerted to listen to the first two speakers and realize the level at which they had pitched their remarks. So, again, please forgive me, but let's see what we can get out of it. The subject then is inelastic electron scattering with unpolarized particles. These, of course, were the experiments that were first performed in the famous MIT-SLAC collaborations now several years old. The experimental idea is very crudely symbolized in Fig. 1.

This is an experiment in which an electron is scattered under conditions where the properties of the incident and the scattered electron are determined. That is, the energies and angles are known. As a result, an

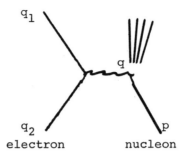

q_1

q

q_2

electron   nucleon

Figure 1.

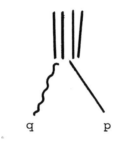

q p

Figure 2.

electromagnetic field in the form of a virtual photon is set up, and through the control of the electron one specifies the energy and the momentum of this virtual photon. Virtual because the energy–momentum relations are not at all that of a real photon. The photon then interacts with a nucleon (experiments have been performed on both neutron and proton) as a result of which, at very high energies, large numbers of particles are emitted. The experiments I'm going to deal with here exert no control over which particles appear in the final state, so these are total cross sections in a certain sense. In fact, as I indicate in Fig. 1, the electrons are there simply to create the "photon", the virtual photon. The physical process then is one of absorption of the photon by the nucleon to produce all possible multi-nucleon states (Fig. 2).

Now, since I'm not going to use models, obviously I have to replace them with something else, and a certain amount of formalism, which is of a rather elementary nature, is required. It is the mathematical expression of quite simple physical ideas, but may be rather strange at first encounter. I can only apologize for that. But the idea is that this picture (Fig. 2), if it represents

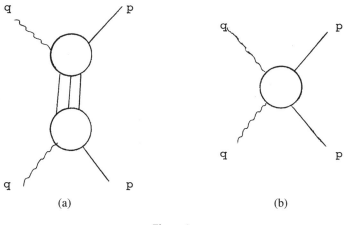

(a)   (b)

Figure 3.

anything, is a sort of shorthand for the quantum mechanical description of the process in terms of a probability amplitude — the amplitude for photon absorption by the nucleon to produce some multi-nucleon state. However, what one is really interested in is the total probability, the absolute square of that amplitude summed over all possible things that can happen. Think of the probability amplitude for the emission process multiplied by its complex conjugate. In quantum mechanics taking the complex conjugate is to some extent running things backward.

Therefore, this is the picture (Figure 3(a)) for the process of computing not the amplitude but the probability, in which we have the particles coming together to form the general multi-particle state and then coming back out again to reform the initial state. And one can also recognize this process as Compton scattering — photon plus nucleon combining to again form a photon and a nucleon (Figure 3(b)). It is forward Compton scattering because the final state is the same as the initial state, that is, the final-photon is identical with the initial photon.

What I'm saying here is perfectly standard in all ways of doing this. I am however, removing the operator language that is customarily used to express it. So we are interested in forward Compton scattering, and, through its description, we will have an account of the virtual photon absorption that occurs in our original problem. Let's then begin to see how we can quantitatively characterize this forward Compton scattering. We are dealing with photons or electromagnetic excitations and the important physical

Table 1.   Gauge invariant combinations.

(1) $-\dfrac{1}{2}F^{\mu\nu}(-q)\,F_{\mu\nu}(q) \equiv |E|^2 - |H|^2.$

(2) $-\dfrac{1}{2}F^{\mu\nu}(-q)\underbrace{\left(g_{\nu\lambda}+\dfrac{p_\nu p_\lambda}{M^2}\right)}_{\text{projects} \perp \text{to } p}F^{\lambda\kappa}(q)\left(g_{\kappa\mu}+\dfrac{p_\kappa p_\mu}{M^2}\right)\underset{\substack{\text{(in the } p \text{ rest frame)}}}{\equiv |H|^2}.$

requirements is that of gauge invariance. And so you will see that I list various gauge invariant combinations of electromagnetic fields in Table 1. The emphasis is on the electromagnetic fields, because since the nucleons are unpolarized, they carry no kinematically important properties except their momentum. No further reference to their characteristics is involved. So here I've listed particular ways, the ways I consider especially significant, of constructing gauge invariant combinations:

Again, forgive me for the technical notation, but $F_{\mu\nu}(q)$ is the electromagnetic tensor that is associated with the incoming electromagnetic field of associated momentum q. In $F_{\mu\nu}(-q)$ we have something similar for the outgoing electromagnetic field. It takes away momentum instead of supplying it. And in (1) of Table 1 we see the very familiar gauge invariant scalar combination of electromagnetic fields which I've also written out in ordinary three-dimensional notation. That's not the only way of forming such combinations, however, it makes no reference to the nucleon — the proton, for example. But the momentum of the proton may figure as well and in fact there is just one other combination; it is (2). I've written out this elaborate thing but it has a very simple meaning. Namely, the proton momentum is being used here through the particular combination which projects perpendicularly to the momentum. If, in fact, you think of things in the laboratory system where the nucleon is at rest, where the nucleon momentum has only a time component — its energy, then this is simply the tensor to pick out the spatial components. And so this elaborate structure evaluated in this natural coordinate system, the p rest frame, is simply the absolute square of the magnetic field. Between these two, then, we have the square of the magnetic field and the square of the electric field, the two independent measures of the character of the electromagnetic field. We are free to control them to a large extent because of the fact that, through

Table 2.   Kinematical scalar variable.

---

(1) $p^2 = -m^2$, m is nucleon mass.

(2) $q^2 > 0$,   (space-like) momentum transfer.

(3) $v = \dfrac{-qp}{m}$,   energy loss of electron in p rest frame.

---

electron scattering, the energy and the momentum of the virtual photon are, to some degree, independently variable.

Now I also remind you of the fact that in any description of such a quantum mechanical process, a fundamental role is played by the various kinematical variables: the four-dimensional momenta and in particular the scalar combinations that can be made from them. And here they are.

Of course, the square of the nucleon momentum is not a variable quantity since it is determined by the mass of the nucleon. The minus sign here is simply my convention that time-like quantities are of negative square and space-like quantities are positive square. In this particular arrangement in which the electron is deflected, the momentum that it transfers, that it creates, is in fact space-like. Accordingly, q, the squared momentum associated with the virtual photon, is positive. This is the invariant measure of the momentum transfer from the electrons to the nucleons. Then, the third combination is the product of the two vectors, the photon momentum and the nucleon momentum. This, when evaluated in the p rest frame, is the nucleon mass multiplied by the same component of q, and therefore, proportional to the energy loss of the electron — one of the experimentally controllable variables.

I now write down the forward Compton scattering amplitude, putting together these various ingredients:

$$1 + iV \, d\omega_p \, 4e^2 \, A^\mu(-q) \left[\sum_{a=1}^{2} T_{a\mu\nu} H_a(q^2, pq) \right] A^\nu(q)$$

where

$$d\omega_p = (2\pi)^{-3} d\vec{p}/2p^\circ, \quad p^\circ = (\vec{p}^2 + m^2)^{1/2}. \tag{1}$$

Unity refers to the situation holding almost everywhere in which the particles simply miss each other, and the probability amplitude for that is just one. The second term describes what happens when the particles do interact. It has the following ingredients: (1) i is there because a scattering amplitude tends typically to be out of phase with non-scattering amplitudes. (2) V is the four-dimensional interaction volume. The two beams of particles, the photons and the nucleons, are focused on each other. They interact for a certain volume of space and for a certain time and, there is a four-dimensional interaction volume to represent that. (3) $d\omega_p$ refers to the nucleon, namely, it is an invariant measure of the small range of momentum space in which the nucleon is to be found. (4) e is the electron charge in rationalized units, where $e^2/4\pi$ is the fine structure constant, e^2 is therefore present because the electromagnetic field is going to act twice. (5) There are two electromagnetic fields. The field of the photon that comes in, and the field of the photon that goes out. You recall I wrote gauge invariant forms explicitly in terms of the field strengths. Now I make it a little more explicit in terms of the vector potentials from which those field strengths are derived, and so, correspondingly there are two tensor structures which contain the photon momentum and the nucleon momentum. Here they are:

$$T_{1\mu\nu} = m^2(g_{\mu\nu}q^2 - q_\mu q_\nu), \quad q_\mu T_{1\mu\nu} = 0; \qquad (2a)$$

$$T_{2\mu\nu} = q^2 p_\mu p_\nu - qp(q_\mu p_\nu + p_\mu p_\nu)$$
$$+ (qp)^2 g_{\mu\nu} + m^2(q^2 g_{\mu\nu} - q_\mu q_\nu);$$

$$q_\mu T_{2\mu\nu} = p_\nu T_{2\mu\nu} = 0. \qquad (2b)$$

I trust that you still carry enough of the field strength idea with you to realize that $T_{1\mu\nu}$ is related to the product of the field strengths. The q's are the derivatives that act upon the vector potentials to make the field strengths, and a factor of m^2 is inserted to give this a convenient set of dimensions. These things have the dimension of the fourth power in momentum. And the fact that this all came from a gauge invariant combination is verified by multiplying the photon momentum into this tensor, yielding zero. The second one looks much more complicated. But it is just that elaborate combination of projection tensors and field strengths designed to pick out the square of the magnetic field in the laboratory system. Its intrinsic meaning is very elementary indeed. As you see, it expresses a gauge invariant combination since it is zero when multiplied by the photon momentum. It also satisfied the second requirement to pick out things that are orthogonal to the particle momentum. These are the basic ingredients, with each tensor multiplied by an arbitrary function $H_{1,2}$ of the scalar variables.

These tensors have two general properties. They are both symmetrical in the vector indices of the electromagnetic fields. Secondly, they are both even functions in q. This is related to a property which is generally called crossing. You notice that a particular momentum q appears in the forward scattering amplitude. But actually this is just a shorthand, because if we wrote down the forward scattering amplitude for an arbitrary electromagnetic field we would really be summing over all possible momentum transfers. In particular, along with q, $-q$ would occur everywhere. As an expression of what is basically Bose–Einstein statistics, there must be a symmetry between these. In other words, if you replace q with $-q$, everything must look the same. Replacing q by $-q$ interchanges these two vector potentials. You should also interchange their indices but that's not necessary here because everything is symmetrical. The whole structure must have a crossing symmetry, which is to say that it must be even under the replacement of q by $-q$. That's true of the tensors. It must, therefore, be true equally well of the functions H_1 and H_2. So the H's must be even functions in qp. That's just something to keep in mind.

Now this is a very general description. Actually its much simpler to consider, not the general electromagnetic field, but specific polarizations. The possibility of varying the polarization depends on the freedom to vary the electromagnetic field which in turn is tied to the possibility of choosing different angles of deflection and different energy losses in the actual electron scattering. There are in fact just two kinds of polarizations in this situation.

Polarization

$$L : A_\mu(q) = n_\mu \quad \text{(unit time-like vector in } q-p \text{ plane)} \qquad (3a)$$

where

$$n_\mu = \frac{q^2 p_\mu - (q \cdot p)q_\mu}{\sqrt{m^2 q^2 + (qp)^2 q^2}}$$

$$T : A_\mu(q) = m_\mu \quad \text{(unit space-like vector } \perp \text{ to } q \text{ and } p)$$

The one that I've called longitudinal is often called scalar. For that situation the vector potential is a unit time-like vector multiplied by a scalar measure of the field strength. The unit time-like vector is in the $q-p$ plane and I have written its actual construction. The second kind of polarization is called transverse. It is constructed from any unit space-like vector. This is a space-like vector perpendicular to q and perpendicular to p. The details don't matter. These properties characterize it, together with a measure of the strength of the electromagnetic field. And now I write the amplitudes, namely the forward scattering amplitudes, specialized to longitudinal polarization and transverse polarization.

$$L : 1 + iV\, d\omega_p\, 4e^2\, A(-q)A(q)\, m^2 q^2 H_1 \tag{4a}$$

$$T : 1 + iV\, d\omega_p\, 4e^2\, A(-q)A(q)\, m^2\, [-q^2 H_1 + (q^2 + v^2) H_2] \tag{4b}$$

where

$$v = qp/m.$$

We note that for longitudinal polarization only the function of H_1 is picked out because the tensor associated with H_2 is perpendicular to p. When we look at transverse polarizations, both H_1 and H_2 enter in a particular combination. You see that the scattering amplitudes involve the following general elements: interaction volume, density of the nucleon, e^2 as an intrinsic measure of the strength of electromagnetism, measure of the intensity of the electromagnetic field and then two functions of the independently variable quantities characterizing the experiment, q^2 and qp.

Now, what we're interested in, of course, is not probability amplitudes but probabilities. And so those two probability amplitudes that I just wrote down are to be squared to give the probability of forward scattering. However, from the fact that the total probability of something happening is one, we get as the complement of the probability of nothing happening (namely, forward scattering) the total probability of something happening. The probability of scattering forward is one minus the probability of some deflection, of any kind whatever. So when we take the absolute square of the forward scattering amplitude, we can infer the total transition probability. All of this is standard optics, and such reasoning is called the Optical Theorem. The interference between the scattering amplitude of unity and the scattering amplitude that starts out of phase, with an i in front, gives the transition probabilities in terms of the imaginary parts of those two H functions. Now what we are interested in, and in fact what is physically measured, are, cross sections. So I want to go directly to the cross sections and work with the cross sections as the quantities of physical interest. Most of the literature of this subject is expressed in terms of other quantities, of so-called structure functions. If you look at the experimental literature, you will find that the cross sections are measured first and then the structure functions are computed from them. The cross section is generally defined in the following way: The total transition probability per unit interaction volume, per unit flux. I call it invariant flux, because it is an invariant measure of the densities and the relative velocities of the two beams of particles with respect to each other. Here are the explicit expressions for the cross sections — longitudinal cross section, transverse cross section.

Cross sections

$$\sigma(\text{total cross section}) = \frac{\text{Total transition probability}}{V(\text{invariant volume}) \, F(\text{invariant flux})}$$

$$F = d\omega_p \, A(-q) A(q) 4\sqrt{m^2 q^2 + (qp)^2},$$

$$\sigma_L = \frac{8\pi\alpha}{m\nu} \frac{1}{\sqrt{1 + (q/\nu)^2}} m^2 q^2 \, \text{Im} \, H_1, \tag{5a}$$

$$\sigma_T = \frac{8\pi\alpha}{m\nu} \frac{1}{\sqrt{1 + (q/\nu)^2}} m^2$$
$$\times [(q^2 + \nu^2) \, \text{Im} \, H_2 - q^2 \, \text{Im} \, H_1], \tag{5b}$$

$$\sigma = \sigma_L + \sigma_T = \frac{8\pi\alpha}{m_\nu} \frac{1}{\sqrt{1 + (q/\nu)^2}} m^2 (q^2 + \nu^2) \text{Im} H_2, \tag{5c}$$

$$\sigma_L/\sigma = (q^2/q^2 + \nu^2) \, \text{Im} \, H_1/\text{Im} \, H_2.$$

They involve kinematical factors and the imaginary parts of the two basic invariant functions, H_1 and H_2. By the way, I also note what the "total" cross section is, that is to say, the sum of longitudinal and transverse cross section. And you will notice the simplifying feature that while the longitudinal cross section involves only H_1, the total cross section involves only H_2. So each of these functions plays a determining role with regard to a particular kind of physical process as characterized by a certain cross section. Also, if you look at these expressions and recall the dimensions of the various quantities, i.e.

m and v are both masses or energies, you see the factor $(mv)^{-1}$ is designed to produce the dimensions of cross sections. Since you will also find four powers of momenta in front of the H functions, both of these must be inverse fourth powers of momenta, which is something to keep in mind. Finally, I've written a useful thing for comparison with experiments, the ratio of the longitudinal cross section to the total cross section, which you see involves a kinematical factor and the ratio of the imaginary parts of H_1 to H_2. Instead of this, the longitudinal/transverse is often used. But since the ratio in practice is small, σ_L/σ is just as good a measure and I prefer to use it because it is simpler.

Now let me point out to you what happens in a related but independent physical situation in which we deal with real photons. The physical process then is just photo-absorption, in which a photon beam falls on a nucleon and we measure the total cross section.

Real photon cross section $(q^2 = 0)$

$$\sigma_L(q^2) = 0 \tag{6a}$$

$$\sigma_\gamma = \sigma_T = \frac{8\pi\alpha}{m^2} m^3 v \, \text{Im} \, H_2(q^2 = 0, qp = -mv) \tag{6b}$$

$$\text{Experiment:} \left[\begin{array}{l} (v/m \gg 1)\, \sigma_\gamma(n)\, \sigma_\gamma(p) \to 1 \\ m^3 v \, \text{Im} \, H_2(q^2 = 0) \cong 1 \end{array} \right.$$

$$\text{Approx. description:}\ \sigma_\gamma(n)/\sigma_\gamma(p) \cong 1 - \frac{1}{4}\sqrt{\frac{m}{v}}. \tag{6c}$$

Of course, a real photon possesses only transverse polarization, and so one would expect that the longitudinal cross section disappears. I denote the total cross section, which is the only thing of interest, by σ_γ. I have written it in terms of a unit, $8\pi\alpha/m^2$ which is $\simeq 80$ micorbarns. Multiplying this unit is the invariant function H_2 (its imaginary part) evaluated for the photon situation. So of the two variables in general, q^2 is zero here and there is only one variable, which is v. Now let me remind you of some experimental facts, because, if I am not to use models, I must, of course, put in something. That something consists of utilizing other experimental bits of information which border on but are not themselves deep inelastic scattering. Of course, all workers in this field make use of related experimental facts. The only question is whether you can handle the facts directly, or whether you always have to go through the intermediary of a model. I simply want to cut out the middle man, the model, and work directly with the experimental facts. At large photon energies compared to the nucleon rest mass, the proton and neutron cross sections seem to approach a common limit. They are slowly, asymptotically, approaching a constant at the energies available. A reasonable extrapolation gives a value of $m^3 \nu \text{Im} H_2$ or about one, which is why I've written it as I have. Actually, it's 1.2, but we will round it off to 1. Secondly, we are going to be interested in the manner in which the neutron and the proton cross sections approach equality. An approximate fit to the data is given by Eq. (6c). I'll remind you of this again. I want to use this as a basic experimental input.

So far we have had just general formalism and a few experimental facts upon which to draw. Now, we need a little more than that, and here is where we get rather more specific. Namely, under the heading of double spectral forms, I will describe to you the general theoretical framework in which all of this is going to be imbedded. In order to appreciate what I am going to say, think, not of the actual experimental arrangement, but a special space-time arrangement.

Double spectral forms

Here, in other words, are momenta not necessarily having the values that are of physical interest. We will get to that later by a process of extrapolation. Think of two nucleon fields and two electromagnetic fields in a particular arrangement. The nucleon field injects some momentum not necessarily

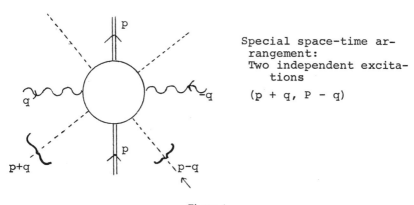

Figure 4.

related to the nucleon mass. Then, in addition, there are, from the sides, spacelike insertions of photon momentum q, and photon momentum $-q$. The net result, for the total momentum injected is p, and that goes out in the form of the nucleon. A nucleon in the general sense, that is, a virtual nucleon. So here we have a coupling of four fields, the same four fields we're interested in but under more general space-time or kinematical conditions. The main point of thinking of it this way is that we can usefully view it in another manner which I have already suggested here. Namely, by thinking of these things as being somewhat variable in their space-time arrangements, you can regard the momentum p and the momentum q, combined together as a total excitation of momentum $p + q$, running through the system in a time-like direction. And then quite independently of that, we can alternatively think of p and the other electromagnetic field with momentum $-q$ producing an excitation of momentum $p - q$. Our ability, to move these around in various ways, to control the space-time arrangement, gives us the possibility of considering these two independent excitations of momenta $p + q$, $p - q$. Well, p and q or their various squares are the basic independent variables. I am just directing your attention here through a particular space-time arrangement, and a certain way of thinking of things, to the importance of the particular combinations $p + q$, $p - q$. The result is to consider an excitation with total momentum $p + q$ running through the system and setting up states of various masses, which I will call M_+. Independently a momentum $p - q$ sets up states of various masses, which I will call M_-. One is lead to superimpose the two and construct the two functions of the variables q^2 and qp in terms of $p + q$ and $p - q$:

$$H_a(q^2, qp)$$
$$= \int \frac{dM_+^2}{M_+^2} \int \frac{dM_-^2}{M_-^2} \frac{2h_a(M_+^2, M_-^2)}{[(p + q)^2 + M_+^2 - i0][(p - q)^2 + M_-^2 - i0]}.$$
$$(7)$$

The denominators are standard propagation functions of these particular momentum excitations. Multiplying these are arbitrary spectral weight functions that measure the probability of exciting the general spectrum. They are dimensionless (recall the inverse fourth power of momentum for the dimensions of H_a). I also point out to you the requirement of crossing symmetry. That is, Eq. (7) should be symmetrical between q and $-q$ or symmetrical between M_+ and M_-. So here is the general idea of considering related physical situations, insisting on their generality, and then extrapolating to the situation of interest.

It's always possible to add single spectral forms, things that depend upon only one of these variables at a time. But I will use the assumption that there are no single spectral forms, and I merely point to the fact there seems to be no need for them. There are no contradictions, no difficulties, and there are comparisons with experiment and rather special situations that motivate us to exclude single spectral forms for this particular situation. Now what we are interested in, is the imaginary part, and here I've indicated to you the possibility of picking it out:

$$
\frac{1}{\pi} \text{Im} \, H_a = \int \frac{dM_+^2}{M_+^2} \int \frac{dM_-^2}{M_-^2} \frac{\delta[(p+q)^2 + M_+^2] \, 2h_a}{(p-q)^2 + M_-^2}
$$

$$
= \int \frac{dM_+^2}{M_+^2} \int \frac{dM_-^2}{M_-^2} \frac{\delta(q^2 - 2Mv - m^2 + M_+^2) h_a}{q^2 + \frac{1}{2}(M_-^2 + M_+^2) - m^2}, \quad (8a)
$$

since

$$
\text{Im} \, \frac{1}{x - i0} = \pi \delta(x). \quad (8b)
$$

Notice that in Eq. (7) both denominators possess an $i0$, which is the characteristic reference to propagation. So it is this imaginary part that contributes, that produces a delta function describing the fact that an injection of momentum $p + q$ produces a certain mass, M_+. Now forgive me if I ask you to take on faith quite elementary manipulations which the running clock prevents me from doing for you. It's all quite direct. I've rewritten the delta function in terms of the variables q^2 and v; the denominator is also written conveniently in terms of q^2. So we isolate in the numerator to some extent the energy input and in the denominator the momentum transfer. Then, just as a mathematical device to separate even further the interesting variable, which is q^2, from the details of the spectrum, I take this denominator and I put it in an exponential by introducing a new integration variable:

$$
\frac{1}{q^2 + \frac{1}{2}(M_- + M_+)^2 - m^2}
$$

$$
= \int_0^\infty \frac{d\zeta}{M_+^2} e^{-(q^2/M_+^2)\zeta} e^{-[(M_+^2 + M_-^2 - 2m^2)/2M_+^2]\zeta}, \quad (9a)
$$

$$
h_a \left(\zeta, \frac{m^2}{M_+^2} \right) \equiv \pi \int \frac{dM_-^2}{M_-^2} e^{\frac{-[M_+^2 + M_-^2 - 2m^2]\zeta}{2M_+^2}} h_a(M_+^2, M_-^2) \quad (9b)
$$

leading to:

$$\text{Im } H_a = \frac{1}{(M_+^2)^2} \int_0^\infty d\zeta e^{-(q^2/M_+^2)\zeta} h_a \left(\zeta, \frac{M^2}{M_+^2} \right), \qquad (9c)$$

where:

$$M_+^2 = M^2 + 2mv - q^2.$$

Now, you will notice M_+ is fixed by the energy conditions. For a given energy transfer, a given energy loss, you excite a certain level of mass. So M_+ is fixed but we integrate over all M_-. And here in Eq. (9b) we are integrating over M_-, to give us a new function, the one really of interest, which depends on the dimensional parameter ζ that I put in to get an exponential form. And I've written in (9b) the fact that the dimensionless function of M_+^2 must be related to some characteristic mass and I've taken the nucleon mass just as a convenient characteristic unit. Of course, I mean it only as an order of magnitude. The nucleon mass is a little less than 1 GeV, the low lying resonance structure is of order of 1 GeV. This is just an indication of an order of magnitude. I'll come back to this immediately. And here summarized in Eq. (9c) is what we have reached.

A little bit of physics now enters through the idea that, at the experimental energies that are presently being explored, and specifically at the energies that were involved in the initial MIT-SLAC experiments, there is no reference yet to a higher mass scale beyond that of a typical nucleon mass (1 GeV). In other words, all the phenomena that occur are understood in terms of the creation of more and more pions, nucleons, kaons. Something characteristically new has not yet come in; we have not yet reached the threshold of totally new physical phenomena. When we do, we will then have to introduce a new characteristic mass ratio. Namely, the mass characterizing those new phenomena, whatever they may be, 10 GeV, 100 GeV (that lies open), relative to the variable mass M^2. So we are dealing with a situation which, relative to possible new phenomena, is still at low energy. That fact is inserted in our very elementary description of the observed properties of the high energy experiments by the assumption that the typical order of magnitude of m is set by the nucleon mass. We are not dealing with "100" nucleon mass units anywhere in the description as yet. That possibility is left open for a more general description.

Kinematics:

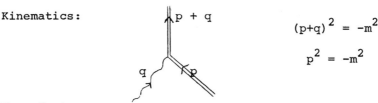

$$(p+q)^2 = -m^2$$

$$p^2 = -m^2$$

Form factors:

$$F_1 \text{(charge)}, \quad F_2 \text{(anomalous mag. mom.)}$$

Figure 5.

I must now turn to some other experimental facts that have to be used. Among all the things that can happen is purely elastic scattering. In other words, the virtual photon can be absorbed by the nucleon to produce just a nucleon with a momentum $(p + q)$. And that the most familiar thing of all, elastic scattering.

Elastic scattering

This is described by two characteristic form factors for the charge and anomalous magnetic moment. It is well known that for elastic scattering the important combinations are given in terms of alternative form factors, namely, the electric and magnetic form factors.

Important combinations

$$G_E = F_1 - (q^2/4m^2)F_2,$$
$$G_M = F_1 + F_2. \tag{10a}$$

Dipole Formula:

$$G_{E,M} \cong 1/(1 + q^2/m_0^2)^2, \quad m_0 \cong 0.9\,m. \tag{10b}$$

I remind you that the form factors G_E and G_M are, for reasons that no one really understands, approximately represented by the so-called dipole formula. Here, the characteristic unit of mass is essentially the nucleon mass, actually 9/10 the nucleon mass. The form factors fall very rapidly with momentum transfers that are large in comparison to the nucleon mass — roughly as the inverse fourth power of the momentum transfer. That need not hold forever, but it seems to be approximately true in the areas of present interest. Now it should come as no surprise that the elastic scattering cross sections can be written in terms of these form factors G_E and G_M. And here is the expression in the framework of Eq. (5):

$$m^2 q^2 \frac{1}{\pi} \operatorname{Im} H_{1,2} = \delta\left(\frac{M_+^2}{m^2} - 1\right) \left\{ G_E^2, \frac{G_E^2 + (q^2/4m^2)G_M^2}{1 + q^2/4m^2} \right\}. \tag{11}$$

This result is standard. But we certainly can express it within the double spectral formalism which covers everything including elastic scattering. For this specific process only the nucleon mass is excited, and so there is a corresponding delta function. This is an independent calculation where we specialize our general space-time analysis to the specific problem of elastic scattering.

Double spectral form (elastic scattering)

$$\operatorname{Im} H_a = \frac{1}{(M_+^2)^2} \int_0^\infty d\zeta\, e^{-(q^2/M_+^2)\zeta} h_a(\zeta, m^2/M_+^2)$$

(1) $h_a(M_+^2, M_-^2) = \delta\left(\dfrac{M_+^2}{m^2} - 1\right)\left[h_a\left(\dfrac{M_-^2}{m^2} - 1\right) + h_a\left(\dfrac{M_-^2}{m^2}\right) \right]$

(2) $h_a(\zeta, m^2/M_+^2) = \pi\delta\left(\dfrac{M_+^2}{m^2} - 1\right) h_a(\zeta)$

where

$$h_a(\zeta) = h_a + \int_{>m^2} \frac{dm^2}{M_-^2} e^{-\frac{(M_-^2 - M^2)\zeta}{2m^2}} h_a(M_-^2/m^2)$$

so that

$$\operatorname{Im} H_a = \frac{\pi}{(m^2)^2} \delta\left(\frac{M_+^2}{m^2} - 1\right)$$

$$\times \left[\frac{m^2}{q^2} h_a + m^2 \int_{>m^2} \frac{dM_-^2}{M_-^2} \frac{h_a(M_-^2/m^2)}{q^2 + \frac{1}{2}(M_-^2 - m^2)} \right]. \tag{11b}$$

You see, we have now an alternate expression for the functions of interest, $\operatorname{Im} H_1$ and $\operatorname{Im} H_2$. The assumption that only the nucleon mass is excited in elastic scattering is explicitly shown here. Furthermore, we have separated the continuum contribution from the single nucleon contribution in the M_- integration appearing in the dimensionless weight functions. Now, a comparison with the previous results equates known form factors on the left to rather particular functions of the spectral weight functions on the right:

$$G_{E}^{2}, \frac{G_{E}^{2} + (q^{2}/4m^{2})G_{M}^{2}}{1 + q^{2}/4m^{2}} = h_{a} + q^{2}\int_{>m^{2}} \frac{dM_{-}^{2}}{M^{2}} \frac{h_{a}(M_{-}^{2}/m^{2})}{q^{2} + \frac{1}{2}(M_{-}^{2} - m^{2})}. \quad (11c)$$

A simple integration by parts yields this second spectral representation of the form factors. Let me point out something to you. If one takes the characteristic momentum dependence in the denominator of the spectral integral and identifies it roughly with the momentum dependence of the Dipole formula, so that m_{o}^{2} equals an average of $1/2(M_{-}^{2} - m^{2})$, you get $M_{-} = 1.5\,\mathrm{GeV}$ as the average excitation involved in producing the elastic form factors. This is considered the resonance region. The large q^{2} behavior of these form factors determines the small ς behavior of the spectral weight functions. It is not difficult to derive this dependence if we again make use of the Dipole formula:

$$\int_{0}^{\infty} d\zeta\, e^{\frac{-q^{2}}{m^{2}}\zeta} h_{a}'(\zeta)\bigg|_{q^{2}/m^{2}\gg 1} = \lim_{q^{2}\to\text{large}} \left\{ G_{E'}^{2} \frac{G_{E}^{2} + (q^{2}/4m^{2})G_{M}^{2}}{1 + (q^{2}/4m^{2})} \right\}$$

$$\cong \left(\frac{m_{0}^{2}}{q^{2}}\right)^{4}.$$

Therefore,

$$\zeta \ll 1, \quad h_{a}'(\zeta) \sim \zeta^{3}. \quad (12)$$

When we go to higher excitations, we certainly expect that the delta function spike, which is the very sharp nucleon excitation, has to disappear. We know experimentally that the resonances quickly blend into each other and form a rather smooth pattern. So I suggest that the delta function of elastic scattering is replaced by a smooth function as the level of excitation gets higher and higher, while the ς dependence of Eq. (12) is perhaps generally valid since experimental evidence in the resonance region indicates the same pattern of decrease for large q^{2}.

I am putting in, of course, a lot of qualitative experimental facts. Let's continue and see where it all leads. Here, we are going to investigate the behavior of the important spectral weight functions in the limit as the excitation level becomes very large compared to the nucleon mass. A reasonable assumption, when the variable m^2/M_+^2 is small, would be that the weight functions only depend on the variable ζ. Physical conditions should not be sensitive to small things. However, this can't be completely true. I say that not out of any general *a priori* convictions, but by looking at what happens with real photons. For real photons in the diffraction region, by which I mean q^2 zero and v/m large, the cross section approaches a constant. This is characteristic of diffraction phenomena. I remind you that at large energies the measure of the cross section, taking away the convenient unit $8\pi\alpha/m^2$, is approximately one. I also recognize that under these conditions the quantity called $2mv$, for real photons at large energies, is the square of the mass that is being excited. For this particular situation we can derive a most interesting relationship between the spectral weight function and the ratio m^2/M_+^2.

Here it is.

Diffraction Region: $(q^2 = 0)$

$$v/m \gg 1, \quad 2m \simeq M_+^2$$

$$\begin{bmatrix} (1) & m^3 v \operatorname{Im} H_2(q^2 = 0, mv) \cong 1 \\ (2) & \operatorname{Im} H_2(0, mv) = \dfrac{1}{(M_+^2)^2} \displaystyle\int_0^\infty d\zeta\, h_2\left(\zeta, \dfrac{m^2}{M_+^2}\right). \end{bmatrix}$$

Hence,

$$\int_0^\infty d\zeta h_2\left(\zeta, \frac{m^2}{M_+^2}\right) \cong \frac{2M_+^2}{m^2} \gg 1. \tag{13}$$

It is obviously impossible to ignore the dependence on the variable m^2/M_+^2, because there is m, in the denominator! In other words, if we regarded m as effectively zero because we are working at a much higher level of excitation, we would run into the paradox of having the right side go to infinity. The possibility of neglecting this small variable is correct if ζ is not particularly large. But if ζ is large, there is a sensitivity to this small variable. Here is a situation in which a mass that might ordinarily be considered negligible, the nucleon mass, is, nevertheless, important because it sets the scale of the constant photon cross section. For that reason m cannot be ignored.

We conclude this survey of related phenomena by listing the plausible behavior of the spectral weight functions in various important kinematical regions, namely:

Table 3. Weight functions.

(1)	$\dfrac{m^2}{M_+^2}\zeta \ll 1$	$h_a(\zeta, m^2/M_+^2) \cong \overline{h}_a(\zeta); \quad \zeta \ll 1; \quad \overline{h}_a'(\zeta) \sim \zeta^3.$
(2)	$\dfrac{m^2}{M_+^2}\zeta \gtrsim 1$	$h_2(\zeta, m^2/M_+^2) = e^{\frac{-m^2}{2M_+^2}\zeta}, \quad$ satisfies Eq. (13).
(3)	$\zeta \gg 1$	$\overline{h}_2(\zeta) \to 1, \quad$ joining 1 and 2.

Now, believe it or not, this has all been background to set the stage for the subject of this talk, namely, deep inelastic scattering. Our previous results will serve as important boundary conditions on the spectral weight functions in the scaling region.

Deep inelastic scattering

What is deep inelastic scattering? It is the circumstance in which the energy transfer is large compared to the nucleon mass, and the momentum transfer is large compared to the nucleon mass, but the ratio of the two is any number in excess of one. This ratio, the variable commonly called ω, is given by $2m\nu/q^2$. The range of allowed values for ω, $(1 < \omega <, \infty)$, is determined just from kinematics $(M_+^2 > m^2)$:

Table 4. Kinematics.

(1)	$q^2/m^2 \gg 1, \quad \omega = 2m\nu/q^2 > 1.$
(2)	$\nu^2/q^2 = (\omega/2)^2 \, q^2/m^2 \gg 1.$
(3)	$M_+^2 \cong 2m\nu - q^2 = (\omega - 1)q^2 (> m^2).$

Now, it should come as no surprise when I say that our analysis of deep inelastic scattering will be based on the double spectral formalism, by inserting the appropriate kinematics. Here it is:

$$2mvq^2\,\mathrm{Im}H_a = \frac{\omega}{(\omega-1)^2}\int_0^\infty d\zeta\,e^{-\zeta/(\omega-1)}h_a(\zeta,m^2/M_+^2)$$

$$\rightarrow \frac{\omega}{(\omega-1)^2}\int_0^\infty d\zeta\,e^{-\zeta/(\omega-1)}\overline{h}_a(\zeta) \equiv f_a(\omega),\quad (14a)$$

$$f_a(\omega) = \frac{\omega}{(\omega-1)}\int_0^\infty d\zeta\,e^{-\zeta/(\omega-1)}h_a'(\zeta) \tag{14b}$$

where $m^2/M_+^2 \ll 1$, finite ω.

First, we know from our analysis of the diffractive and resonance regions that at a high level of excitation the spectral weight functions depend only on the variable ζ — provided that ζ is not too large. Such is the case for deep inelastic scattering, provided we restrict ourselves to a region of finite ω. Secondly, a simple integration by parts has been performed giving an alternative form for $f_a(\omega)$. It is evident that as ω approaches one the small ζ behavior of the weight functions is probed. Conversely, as we allow the value of ω to increase the large ζ behavior of the weight functions is probed. The dependence of the spectral weight functions on the variable ζ, however, is already known for these two cases from elastic and diffractive scattering. Inserting our previous results into $f_a(\omega)$, we find:

(1) $\omega \rightarrow 1$, $\zeta \ll 1$; $\overline{h}_a'(\zeta) \simeq \zeta^3$

$$f_a(\omega) = \frac{\omega}{\omega-1}\int_0^\infty d\zeta\,e^{-\zeta/(\omega-1)}\zeta^3 \cong (\omega-1)^3. \tag{15a}$$

(2) $\omega \gg 1$, $\zeta \gg 1$; $\overline{h}_2(\zeta) \rightarrow 1$

$$f_2(\omega) \cong \frac{\omega}{(\omega-1)^2}\int_0^\infty d\zeta\,e^{-\zeta/(\omega-1)} = \frac{\omega}{\omega-1} \cong 1. \tag{15b}$$

From this we infer the deep inelastic behavior of the cross sections:

(1) $\sigma_L/\sigma = [q^2/(q^2+v^2)]\mathrm{Im}H_1/\mathrm{Im}H_2 \rightarrow q^2/v^2\,f_1(\omega)/f_2(\omega)$,

or

$$\sigma_L/\sigma \to \frac{1}{q^2}\left(\frac{2m}{\omega}\right)^2 \frac{f_1(\omega)}{f_2(\omega)} = \frac{m^2}{q^2}\phi(\omega) \ll 1. \tag{16a}$$

$$(2) \quad \sigma = \frac{8\pi\alpha}{m\nu}\frac{1}{\sqrt{1+(q^2/\nu^2)}}m^2(q^2+\nu^2)\,\mathrm{Im}\,H_2 \cong (8\pi\alpha m\nu)\,\mathrm{Im}\,H_2$$

$$= \frac{4\pi\alpha}{q^2}(2m\nu q^2\,\mathrm{Im}\,H_2) \to \frac{4\pi\alpha}{q^2}f_2(\omega). \tag{16b}$$

Now we compare with experimental observation. Deep inelastic scattering is characterized by: (1) A very small ratio of longitudinal to total cross section, (2) A total cross section that depends on a factor $4\pi\alpha/q^2$ times a function f of the single variable ω. It is this behavior that characterizes scaling. That is, the dependence of f is not on the two variables ν and q^2 independently, but on the ratio. (3) When this experiment is done for neutrons and protons (so you have scaling functions for both) it's observed that they approach equality at large ω, but their ratio decreases as ω approaches onc. The experimental limit as ω goes to one remains uncertain. There is good agreement between theory and experiment for the first two scaling properties, where f is given by $f_2(\omega)$. The large and small ω behavior of $f(\omega)$ is correctly described by Eqs. (15a) and (15b). This is encouraging. We still need the ω dependence for the ratio of neutron to proton cross sections in the scaling region. A clue to the ω behavior of σ_n/σ_p is provided by the scattering of real photons. Here, we know that $\sigma_n/\sigma_p \cong 1 - \frac{1}{4}\sqrt{m/\nu}$. Now, for real photons, a large mass excitation can be directly related to the energy transfer by $M_+^2 \cong 2m\nu$. Substituting this gives $\sigma_n/\sigma_p = 1 - \frac{1}{2}\sqrt{\frac{\frac{1}{2}m^2}{M_+^2}}$. This is a very suggestive relationship, for it implies how we can extrapolate this ratio of cross sections to any q^2. For $q^2 > 0$, the factor $e^{-(q^2/M_+^2)\varsigma}$ in Eq. (9c) is joined with the factor $e^{-(\frac{1}{2}m^2/M_+^2)\varsigma}$ of Table 3 to give $e^{-[(q^2+\frac{1}{2}m^2)/M_+^2]\varsigma}$. This indicates the extension of σ_n/σ_p from $q^2 = 0$ according to

$$\sigma_n/\sigma_p \cong 1 - \frac{1}{2}\sqrt{\frac{\frac{1}{2}m^2}{M_+^2}} \to 1 - \frac{1}{2}\sqrt{\frac{q^2+\frac{1}{2}m^2}{M_+^2}} \underset{q^2 \gg \frac{1}{2}m^2}{\cong} 1 - \frac{1}{2}\sqrt{\frac{q^2}{M_+^2}}$$

$$= 1 - \frac{1}{2}\frac{1}{\sqrt{\omega-1}} \tag{17}$$

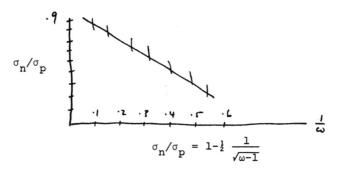

$$\sigma_n/\sigma_p = 1-\tfrac{1}{2}\,\frac{1}{\sqrt{\omega-1}}$$

Figure 6.

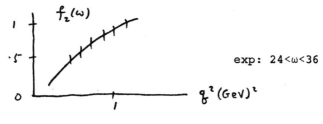

exp: 24<ω<36

Figure 7.

which uses relation (3) of Table 4. This should be valid for $M_+^2 \gg q^2$ or $\omega \gg 1$. The comparison with experiment sketched in Fig. 6 shows that the agreement is quite reasonable for $\omega > 2$.

The same exponential factor $e^{-[(q^2+\frac{1}{2}m^2)/M_+^2]\zeta}$ describe the derivations from scaling. Its introduction in Eq. (15b) gives

$$\omega \gg 1: \quad f_2(\omega) \to \frac{1}{\omega} \int_0^\infty d\zeta\, e^{-\frac{q^2+\frac{1}{2}m^2}{\omega q^-}\zeta} = \frac{q^2}{q^2+\frac{1}{2}m^2} \tag{18}$$

where $\frac{1}{2}m^2$ is actually divided by 1.2 and is $\simeq \frac{1}{3}$ (GeV)^2.
This is a fair fit to experiment:
More generally, we have

$$\mathrm{Im}\, H_2 = \frac{1}{(M_+^2)^2} \int_0^\infty d\zeta\, e^{-\frac{q^2+\frac{1}{2}m^2}{M_+^2}\zeta} \bar{h}_2(\zeta) \tag{19}$$

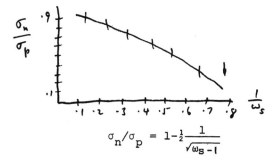

$$\sigma_n/\sigma_p = 1-\tfrac{1}{2}\frac{1}{\sqrt{\omega_s-1}}$$

Figure 8.

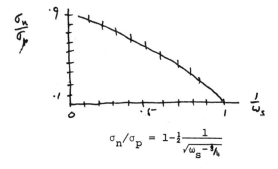

$$\sigma_n/\sigma_p = 1-\tfrac{1}{2}\frac{1}{\sqrt{\omega_s-\tfrac{3}{4}}}$$

Figure 9.

and (16b) is replaced by

$$\sigma \equiv \frac{4\pi\alpha}{q^2}\sqrt{1+\frac{q^2}{\nu^2}}\frac{2m\nu}{2m\nu+\tfrac{3}{2}m^2}\frac{q^2}{q^2+\tfrac{1}{2}m^2}f_2(\omega_S),$$

$$\omega_S = \frac{2m\nu+\tfrac{3}{2}m^2}{q^2+\tfrac{1}{2}m^2} \tag{20}$$

which combines a description of scaling derivations with the indication that ω_s is an improved scaling variable.

If one converts the σ_n/σ_p data to ω_s as a variable (Fig. 7) the good fit is extended down to $\omega_s = 1.5$. As a purely empirical observation, we note that the replacement of ω_s by $\omega_S + \tfrac{1}{4}$, which does no violence to the origin of this formula in large ω, gives an excellent overall fit and, if taken seriously, predict $\sigma_n/\sigma_p \to 0, \omega_s \to 1$.

My conclusion then is that the general characteristics of deep inelastic scattering emerge as reasonable interpolations between the known properties of the low energy resonance region and the properties of the high energy diffraction region, without need for any reference to speculative dynamical models.

I've gone on long enough, but I had prepared a number of other topics. Obviously, we've not reached the end of the game. The next question is to discuss all of this with polarized particles. So far, there have been no such experiments, but one can set up the general background against which they are to be considered. And one can consider electron positron annihilation and face up to the problem that has bothered many theorists these days: why the deviations in scaling seem to be much more pronounced there than they are in deep inelastic electron scattering. But I shall leave the discussion of those further subjects to another time and another place.

4

Life at Low Reynolds Number*

E. M. Purcell

Harvard University, Cambridge, Massachusetts

This is a talk that I would not, I'm afraid, have the nerve to give under any other circumstances. It's a story I've been saving up to tell Viki. Like so many of you here, I've enjoyed from time to time the wonderful experience of exploring with Viki some part of physics, or anything to which we can apply physics. We wander around strictly as amateurs equipped only with some elementary physics, and in the end, it turns out, we improve our understanding of the elementary physics even if we don't throw much light on the other subjects. Now this is that kind of a subject, but I have still another reason for wanting to, as it were, needle Viki with it, because I'm going to talk for a while about viscosity. Viscosity in a liquid will be the dominant theme here and you know Viki's program of explaining everything, including the heights of mountains, with the elementary constants. The viscosity of a liquid is a very tough nut to crack, as he well knows, because when the stuff is cooled by merely forty degrees, its viscosity can change by a factor of a million. I was really amazed by fluid viscosity in the early days of NMR, when it turned out that glycerine was just what we needed to explore the behavior of spin relaxation. And yet if you were a little bug inside the glycerine, looking around, you wouldn't see much change in your surroundings as the glycerine cooled. Viki will say that he can at least

---

*This is a slightly edited transcript of a tape. The figures reproduce transparencies used in the talk. The demonstration involved a tall rectangular transparent vessel of corn syrup, projected by an overhead projector turned on its side. Some essential hand waving cannot be reproduced.

predict the *logarithm* of the viscosity. And that, of course, is correct because the reason viscosity changes is that it's got one of these activation energy things and what he can predict is the order of magnitude of the exponent. But it's more mysterious than that, Viki, because if you look at the Chemical Rubber Handbook table you will find that there is almost no liquid with viscosity much lower than that of water. The viscosities have a big range *but they stop at the same place*, I don't understand that. That's what I'm leaving for him.

Now, I'm going to talk about a world which, as physicists, we almost never think about. The physicist hears about viscosity in high school when he's repeating Millikan's oil drop experiment and he never hears about it again, at least not in what I teach. And Reynolds' number, of course, is something for the engineers. And the *low* Reynolds number regime most engineers aren't even interested in — except possibly chemical engineers, in connection with fluidized beds, a fascinating topic I heard about from a chemical engineer friend at M.I.T.. But I want to take you into the world of very low Reynolds number — a world which is inhabited by the overwhelming majority or the organisms in this room. This world is quite different from the one that we have developed our intuitions in.

I might say what got me into this. To introduce something that will come later, I'm going to talk partly about how micro-organisms swim. That will not, however, turn out to be the only important question about them. I got into this through the work of a former colleague of mine at Harvard, Howard Berg. Berg got his Ph.D. with Norman Ramsey, working on a hydrogen maser, and then he went back into biology which had been his early love, and into cellular physiology. He is now at the University of Colorado at Boulder, and has recently participated in what seems to me one of the most astonishing discoveries about the questions we're going to talk about. So it was partly Howard's work, tracking *E. coli* and finding out this strange thing about them, that got me thinking about this elementary physics stuff.

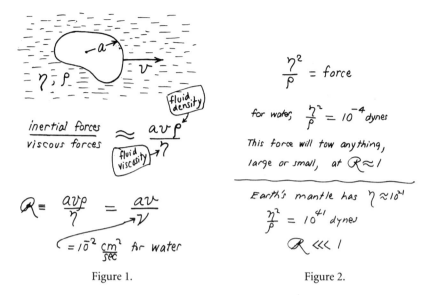

$$\frac{\eta^2}{\rho} = force$$

for water, $\frac{\eta^2}{\rho} = 10^{-4}$ dynes

This force will tow anything, large or small, at $\mathcal{R} \approx 1$

Earth's mantle has $\eta \approx 10^{21}$

$$\frac{\eta^2}{\rho} = 10^{41} \text{ dynes}$$

$\mathcal{R} \lll 1$

$$\frac{inertial\ forces}{viscous\ forces} \approx \frac{a v \rho}{\eta}$$

$$\mathcal{R} = \frac{a v \rho}{\eta} = \frac{a v}{\nu}$$

$\nu = 10^{-2}\ \frac{cm^2}{sec}$ for water

Figure 1.
Figure 2.

Well, here we go. In Fig. 1, you see an object which is moving through a fluid with velocity v. It has dimension a. In Stoke's Law, the object is a sphere, but here it's anything; η and ρ are the viscosity and density of the fluid. The ratio of the inertial forces to the viscous forces, as Osborne Reynolds pointed out slightly less than a hundred years ago, is given by $a v \rho / \eta$ or $a v / \nu$ where ν is called the *kinematic* viscosity. It's easier to remember its dimensions: for water, ν is $\approx 10^{-2}$ cm²/sec. The ratio is called the Reynolds number and when that number is small the viscous forces dominate. Now there is an easy way, which I didn't realize at first, to see who should be interested in small Reynolds numbers. If you take the viscosity and square it and divide by the density, you get a force (Fig. 2). No other dimensions come in at all. η^2 / ρ is a force. For water, since $\eta \approx 10^{-2}$ and $\rho \approx 1$, $\eta^2 / \rho \approx 10^{-4}$ dynes. That is a force that will tow *anything*, large or small, with a Reynolds number of order of magnitude 1. In other words, if you want to tow a submarine with Reynolds number 1 (or strictly speaking, $1/6\pi$ if it's a spherical submarine) tow it with 10^{-4} dynes. So it's clear in this case that you're interested in small Reynolds number if you're interested in *small forces* in the absolute sense. The only other people who are interested in low Reynolds number, although they usually don't have to invoke it, are the geophysicists. The earth's mantle is supposed to have a viscosity of 10^{21} poises. If you now work out η^2 / ρ, the force is 10^{41} dynes. That is more than 10^9 times the gravitational force that

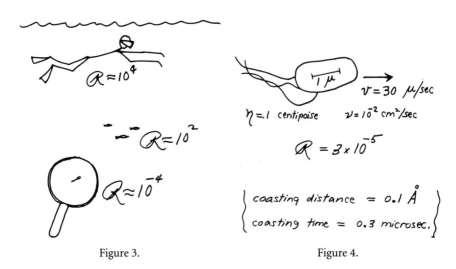

Figure 3.   Figure 4.

half the earth exerts on the other half! So the conclusion is, of course, that in the flow of the mantle of the earth the Reynolds number is *very* small indeed.

Now consider things that move through a liquid (Fig. 3). The Reynolds number for a man swimming in water might be 10^4, if we put in reasonable dimensions. For a goldfish or a tiny guppy it might get down to 10^2. For the animals that we're going to be talking about, as we'll see in a moment, it's about 10^{-4} or 10^{-5}. For these animals inertia is totally irrelevant. We know that $F = ma$, but they could scarcely care less. I'll show you a picture of the real animals in a bit but we are going to be talking about objects which are the order of a micron in size (Fig. 4). That's a micron scale, not a suture, in the animal in Fig. 4. In water where the kinematic viscosity is 10^{-2} cm/sec. these things move around with a typical speed of 30 microns per second. If I have to push that animal to move it, and suddenly I stop pushing, how far will it coast before it slows down? The answer is, about a tenth of an angstrom. And it takes it about three-tenths of a microsecond to slow down. I think this makes clear what low Reynolds number means. Inertia plays no role whatsoever. If you are at low Reynolds number, what you are doing at the moment is entirely determined by the forces that are exerted on you *at that moment*, and by nothing in the past.

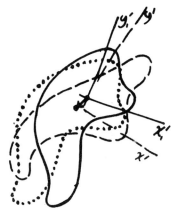

Figure 5.

It helps to imagine under what conditions a man would be swimming at, say, the same Reynolds number as his own sperm. Well, you put him in a swimming pool that is full of molasses, and then you forbid him to move any part of his body faster than one centimeter per minute. Now imagine yourself in that condition: you're under the swimming pool in molasses, and now you can only move like the hands of a clock. If under those ground rules you are able to move a few meters in a couple of weeks, you may qualify as a low Reynolds number swimmer.

I want to talk about swimming at low Reynolds number in a very general way. What does it mean to swim? Well, it means simply that you are in some liquid and are allowed to deform your body in some manner. That's all you can do. Move it around and move it back. Of course, you choose some kind of cyclic deformation because you want to keep swimming, and it doesn't do any good to use a motion that goes to zero asymptotically. You have to keep moving. So, in general, we are interested in cyclic deformations of a body on which there are no external torques or forces except those exerted by the surrounding fluid. In Fig. 5, there is an object which has a shape shown by the solid line; it changes its shape to the dashed contour and then it changes back, when it finally gets back to its original shape, the dotted contour, it has moved over and rotated a little. It has been swimming. When it executed the cycle, a displacement resulted. If it repeats the cycle, it will, of course, effect the same displacement, and in two dimensions we'd see it progressing around some circle. In three dimensions its most general trajectory is a helix

Navier - Stokes :

$$-\nabla p + \eta \nabla^2 \vec{v} = \rho \frac{\partial \vec{v}}{\partial t} + \rho (\vec{v} \times \nabla) \vec{v}$$

If $\mathcal{R} \ll 1$:

Time doesn't matter. The pattern of motion is the same, whether slow or fast, whether forward or backward in time.

The Scallop Theorem

Figure 6.

consisting of a lot of little kinks, each of which is the result of one cycle of shape-change.

There is a very funny thing about motion at low Reynolds number, which is the following. One special kind of swimming motion is what I call a reciprocal motion. That is to say, I change my body into a certain shape and then I go back to the original shape by going through the sequence in reverse. At low Reynolds number, everything reverses just fine. Time, in fact, makes no difference — only configuration. If I change quickly or slowly, the pattern of motion is exactly the same. If you take the Navier–Stokes equation and throw away the inertia terms, all you have left is $\nabla^2 v = p/\eta$ where p is the pressure (Fig. 6). So, if the animal tries to swim by a reciprocal motion, it *can't go anywhere*. Fast or slow, it exactly retraces its trajectory and it's back where it started. A good example of that is a scallop. You know, a scallop opens its shell slowly and closes its shell fast, squirting out water. The moral of this is that the scallop at low Reynolds number is no good. It can't swim because it only has one hinge, and if you have only one degree of freedom in configuration space, you are bound to make a reciprocal motion. There is nothing else you can do. The simplest animal that can swim that way is an animal with two hinges. I don't know whether one exists but

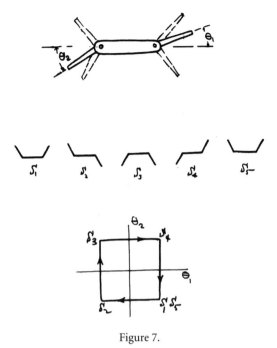

Figure 7.

Fig. 7 shows a hypothetical one. This animal is like a boat with a rudder at both front and back, and nothing else. This animal can swim. All it has to do is go through the sequence of configurations shown, returning to the original one at S5. Its configuration space, of course, is two dimensional with coordinates θ_1, θ_2. The animal is going around a loop in that configuration space, and that enables it to swim. In fact, I worked this one out just for fun and you can prove from symmetry that it goes along the direction shown in the figure. As an exercise for the student, what is it that distinguishes that direction?

You can invent other animals that have no trouble swimming. We had better be able to invent them, since we know they exist. One you might think of first as a physicist, is a torus. I don't know whether there is a toroidal animal, but whatever other physiological problems it might face, it clearly could swim at low Reynolds number (Fig. 8). Another animal might consist of two cells which were stuck together and were able to roll on one another by having some kind of attraction here while releasing there. That thing will "roll" along. I described it once as a combination caterpillar tractor and bicycle built for two, but that isn't the way it really

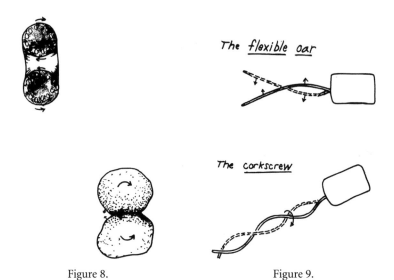

Figure 8.   Figure 9.

works. In the animal kingdom, there are at least two other more common solutions to the problem of swimming at low Reynolds number (Fig. 9). One might be called the flexible oar. You see, you can't row a boat at low Reynolds number in molasses — if you are submerged — because the stiff oars are just reciprocating things. But if the oar is flexible that's not true, because then the oar bends one way during the first half of the stroke and the other during the second half. That's sufficient to elude the theorem that got the scallop. Another method, and the one we'll mainly be talking about, is what I call a corkscrew. If you keep turning it, that, of course is not a reciprocal change in configuration space and that will propel you. At this point, I wish I could persuade you that the direction in which this helical drive will move is *not* obvious. Put yourself back in that swimming pool under the molasses and move around very, very slowly. Your intuitions about *pushing water backwards* are irrelevant. That's not what counts. Now, unfortunately, it turns out that the thing does move the way your naive, untutored and actually incorrect argument would indicate but that's just a pedagogical misfortune that we are always running into.

Well, let's look at some real animals (Fig. 10). This figure I've taken from a paper of Howard Berg's that he sent me. Here are three real swimmers. The one we're going to be talking about most is the famous animal, *Escherichia coli*, at A, which is a very tiny thing. Then there are two larger animals. I've

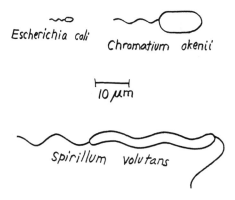

Figure 10.

copied down their Latin names and they may be old friends to some of you here. This thing (*S. volutans*) swims by waving its body as well as its tail and roughly speaking, a spiral wave runs down that tail. The bacterium *E. coli* on the left is about two microns long. The tail is the part that we are interested in. That's the flagellum. Some *E. coli* cells have them coming out the sides; and they may have several, but when they have several they tend to bundle together. Some cells are non-motile and don't have flagella. They live perfectly well, so swimming is not an absolute necessity for this particular animal, but the one in the figure does swim. The flagellum is only about 130 angstroms in diameter. It is much thinner than the cilium which is another very important kind of propulsive machinery. There is a beautiful article on cilia in this month's *Scientific American*.[†] Cilia are about 2000 angstroms in diameter, with a rather elaborate apparatus inside. There's not room for such apparatus inside this flagellum.

For a long time there has been interest in how the flagellum works. Classic work in this field was done around 1951, as I'm sure some of you will remember, by Sir Geoffrey Taylor, the famous fluid dynamicist of Cambridge. One time I heard him give a fascinating lecture at the National Academy. Out of his pocket at the lecture he pulled his working model, a cylindrical body with a helical tail driven by a rubber-band motor inside the body. He had tested it in glycerine. In order to make the tail he hadn't just done the simple thing of having a turning corkscrew, because at that

---

[†] Peter Satir, "How Cilia Move", *Scientific American*, 231, October, 1974, p. 45.

time nearly everyone had persuaded themselves that the tail doesn't rotate, it waves. Because after all, to rotate you'd have to have a rotary joint back at the animal. So he had sheathed the turning helix with rubber tubing anchored to the body. The body had a keel. I remember Sir Geoffrey Taylor saying in his lecture that he was embarrassed that he hadn't put the keel on it first and he'd had to find out that he needed it. There has since been a vast literature on this subject, only a small part of which I'm familiar with. But at that time G.I. Taylor's paper in *Proc. Royal Soc.* could conclude with just three references: Lamb, H., *Hydrodynamics*; Taylor, G.I. (his previous paper); Watson, G.N., *Bessel Functions*. That is called getting in at the ground floor.

To come now to modern times, I want to show a picture of these animals swimming or tracking. This is the work of Howard Berg, and I'll first describe what he did. He started building the apparatus when he was at Harvard. He was interested in studying not the actual mechanics of swimming at all but a much more interesting question, namely, why these things swim and to where they swim. In particular, he wanted to study chemotaxis in *E. coli* seeing how they behave in gradients of nutrients and things like that. So he built a little machine which would track a single bacterium in x, y, z coordinates — just lock onto it optically and track it. He was able then to track one of these objects while it is behaving in its normal manner, possibly subject to the influence of gradients of one thing or another. A great advantage of working with a thing like *E. coli* is that there are so many mutant strains that have been well studied that you can use different mutants for different things. The next picture (Fig. 11) is one of his tracks. It shows a projection on a plane of the track of one bacterium. The little dots are about a tenth of a second apart so that it was actually running along one of the long legs for a second or two and the speed is typically 20–40 microns per second. Notice that it swims for a while and then stops and goes off in some other direction. We'll see later what that might suggest.

A year ago, Howard Berg went out on a limb and wrote a paper in *Nature* in which he argued that, on the basis of available evidence, *E. coli* must swim by *rotating* their flagella, not by waving them. Within the year a very elegant, crucial experiment by Silverman and Simon *at* U.C.-San Diego, showed that this in fact is the case (see references). Their experiment involved a mutant strain of *E. coli* bacteria which don't make flagella at all but only make something called the proximal hook to which the flagella

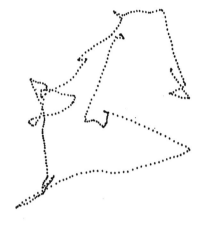

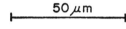

50 μm

Figure 11.

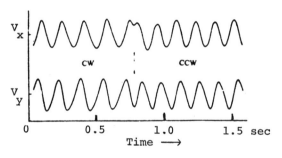

Figure 12.

would have been attached. They found that with anti-hook antibodies they could cause these things to glue together. And once in a while one of the bacteria would have its hook glued to the microscope slide, in which case the whole body rotated at constant angular velocity. And when two hooks glued together, the two bodies counter-rotated, as you would expect. It's a beautiful technique, Howard was ready with his tracker and the next picture (Fig. 12) shows his tracker following the end of one of these tethered *E. coli* cells which is stuck to the microscope slide by antibody at the place where the flagellum should have been. Plotted here are the two velocity components

V_x and V_y. The two velocity components are 90° out of phase. The point being tracked is going in a circle. In the middle of the figure, you see a 90° phase change in one component, a reversal of rotation. They can rotate hundreds of revolutions at constant speed and then turn around and rotate the other way. Evidently the animal actually has a rotary joint, and has a motor inside that's able to drive a flagellum in one direction or the other, a most remarkable piece of machinery.

I got interested in the way a rotating corkscrew can propel something. Let's consider propulsion in one direction only, parallel to the axis of the helix. The helix can translate and it can rotate; you can apply a force to it and a torque. It has a velocity v and an angular velocity Ω. And now remember, at low Reynolds number everything is linear. When everything is linear, you expect to see matrices come in. Force and torque must be related by matrices with constant coefficients, to linear and angular velocity. I call this little two by two matrix the propulsion matrix (Fig. 13). If I knew its elements A, B, C, D, I could then find out how good this rotating helix is for propelling anything.

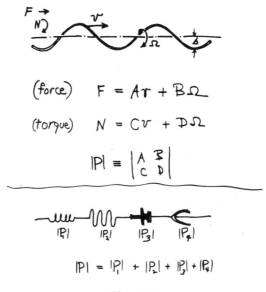

Figure 13.

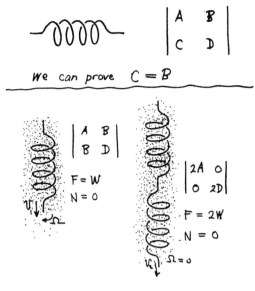

Figure 14.

Well, let's try to go on by making some assumptions. If two corkscrews or other devices on the same shaft are far enough from one another so that their velocity patterns don't interact their propulsive matrices just add. If you allow me that assumption, then there is a very nice way, which I don't have time to explain, of proving that the propulsion matrix must be symmetrical (Fig. 14). So actually the motion is described by only three constants, not four, and they are very easily measured. All you have to do is make a model of this thing and drop in a fluid at low Reynolds number. It doesn't matter whether you have exactly the Reynolds number you are interested in or not, because these constants are independent of that. And so I did that and that's my one demonstration. I thought this series of talks ought to have one experiment and there it is. We're looking through a tank not of glycerine but of corn syrup, which is cheaper, quite uniform, and has a viscosity of about 50 poise or 5,000 times the viscosity of water. The nice part of this is you can just lick the experimental material off your fingers.

Motion at low Reynolds number is very majestic, slow and regular. You'll notice that the model is actually rotating, but rather little. If that were a corkscrew moving through a cork of course, the pattern in projection

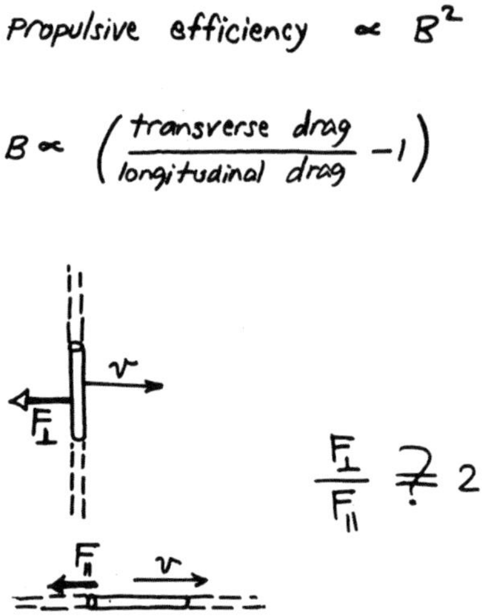

$$Propulsive\ efficiency\ \propto\ B^2$$

$$B \propto \left(\frac{transverse\ drag}{longitudinal\ drag} - 1 \right)$$

$$\frac{F_\perp}{F_\parallel} \not= 2$$

Figure 15.

wouldn't change. It's very very far from that, it's *slipping*, so that it sinks by several wave lengths while it's turning around once. If the matrix were diagonal, the thing would not rotate at all. So all you have to do is just see how much it turns as it sinks and you have got a handle on the off diagonal element. A nice way to determine the other elements is to run two of these animals, one of which is a spiral and the other is two spirals, in series, of opposite handedness. The matrices add and with two spirals of opposite handedness, the propulsion matrix must be diagonal (Fig. 14). That's not going to rotate; it better not.

The propulsive efficiency is more or less proportional to the square of the off diagonal element of the matrix. The off-diagonal element depends on the difference between the drag on a wire moving perpendicular to its length and the drag on a wire moving parallel to its length (Fig. 15). These are supposed to differ in a certain limit by a factor of two. But for the models I've tested that factor is more like 1.5. Since it's that factor minus 1 that counts, that's very bad for efficiency. We thought that if you want something to rotate more while sinking, it would be better not to use a round wire. Something like a slinky ought to be better. I made one and measured its off diagonal elements. Surprise, surprise, it was no better at all! I don't really understand that, because the fluid mechanics of these two situations is not at all simple. I say this not to people who already know

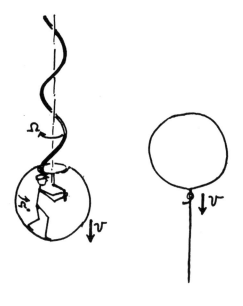

PROPULSIVE EFFICIENCY ≈ 1 %

Figure 16.

that, but to a Stokes Law physicist who would think it might be simple. In each case there is a logarithmic divergence that you have to worry about, and the two are somewhat different in character. So that theoretical ratio of two I referred to is probably not even right.

When you put all this in and calculate the efficiency, you find that it's really rather low even when the various parameters of the model are optimized. For a sphere which is driven by one of these helical propellers (Fig. 16), I will define the efficiency as the ratio of the work that I would have to do just to pull that thing along to what the man inside it turning the crank has to do. And that turns out to be about one percent. I worried about that result for a while and tried to get Howard interested in it. He didn't pay much attention to it, and he shouldn't have, because it turns out that efficiency is really not the primary problem of the animal's motion. We'll see that when we look at the energy requirement. How much power does it take to run one of these things with a one percent efficient propulsion system, at this speed in these conditions? We can work it out very easily. Going 30 microns per second, a one percent efficiency will cost us about 2×10^{-8}

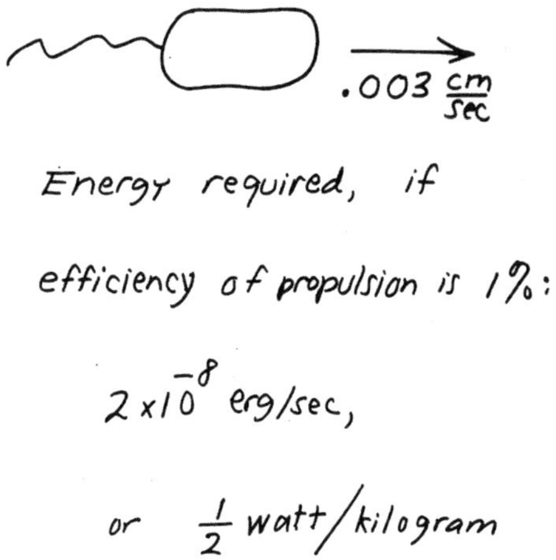

Energy required, if

efficiency of propulsion is 1%:

$$2 \times 10^{-8} \text{ erg/sec},$$

or $\frac{1}{2}$ watt/kilogram

Figure 17.

ergs per second, at the motor. On a per weight basis, that's a half watt per kilogram, which is really not very much. Just moving things around in our transportation system, we use energy at thirty or forty times that rate. This bug runs 24 hours a day and only uses half a watt per kilogram. That's a small fraction of its metabolism and its energy budget. Unlike us, they do not squander their energy budget just moving themselves around. So they don't care whether they have a one percent efficient flagellum or two percent efficient flagellum. It doesn't really make that much difference. They're driving a Datsun in Saudi Arabia.

So the interesting question is not how they swim. Turn anything — if it isn't perfectly symmetrical, you'll swim. If the efficiency is only one percent, who cares. A better way to say it is that the bug can collect, by diffusion through the surrounding medium, enough energetic molecules to keep moving when the concentration of those molecules is 10^{-9} molar. I've now introduced the word diffusion. Diffusion is important because of another very peculiar feature of the world at low Reynolds number and that is, stirring isn't any good. The bug's problem is not its energy supply; its problem is its environment. At low Reynolds number you can't shake off your environment. If you move, you take it along; it only gradually falls behind. We can use elementary physics to look at this in a very simple way. The time for transporting anything a distance L by stirring, is about L divided by the stirring speed v. Whereas, for transport by diffusion, it's

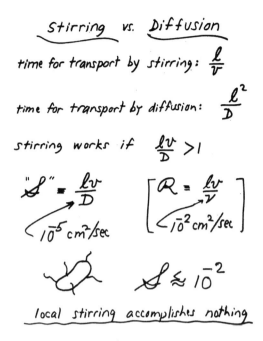

Figure 18.

L^2 divided by D, the diffusion constant. The ratio of those two times is a measure of the effectiveness of stirring versus that of diffusion for any given distance and diffusion constant. I'm sure this ratio has someone's name but I don't know the literature and I don't know whose number that's called. Call it S for *stirring number*.‡ It's just Lv/D. You'll notice by the way that Reynolds number was Lv/ν. ν is the kinematic viscosity in cm²/sec., and D is the diffusion constant in cm²/sec., for whatever it is that we are interested in following — let us say a nutrient molecule in water. Now, in water the diffusion constant is pretty much the same for every reasonably sized molecule, something like 10^{-5} cm²/sec. In the size domain that we're interested in, of micron distances, we find that the stirring number S is 10^{-2}, for the velocities that we are talking about (Fig. 18). In other words, this bug can't do anything by stirring its local surroundings' it might as well wait for things to diffuse, either in or out. The transport of wastes away from the animal and food to the animal evidently is entirely

---

‡I've recently discovered that its official name is the *Sherwood number*, so S is appropriate, after all!

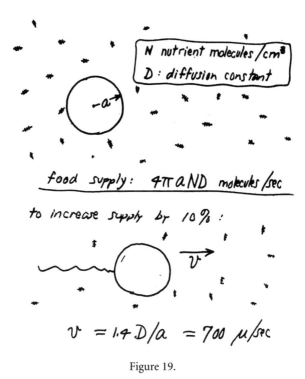

Figure 19.

controlled *locally* by diffusion. You can thrash around a lot, but the fellow who just sits there quietly waiting for stuff to diffuse in will collect just as much.

At one time I thought that the reason the thing swims is that if it swims it can get more stuff, because the medium is full of molecules the bug would like to have. All my instincts as a physicist say you should move if you want to scoop that stuff up. You can easily solve the problem of diffusion in the velocity field represented by the Stokes flow around a sphere — for instance, by a relaxation method. I did so and found out how fast the cell would have to go to increase its food supply. The food supply if it just sits there is $4\pi rD$ molecules/sec, where r is the cell's radius (Fig. 19). To increase its food supply by 10 percent it would have to move at a speed of 700 microns per second, which is 20 times as fast as it can swim. The increased intake varies like the square root of the bug's velocity so the swimming does no good at all in that respect. But what it can do is find places where the food is better or more abundant. That is, it does not move like a cow that is grazing a

to out-swim diffusion:

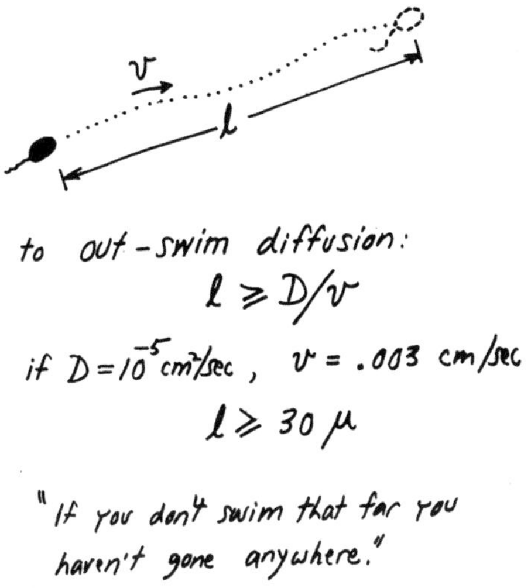

$$\ell \geqslant D/v$$

if $D = 10^{-5} cm^2/sec$, $v = .003 \ cm/sec$

$$\ell \geqslant 30 \ \mu$$

"If you don't swim that far you haven't gone anywhere."

Figure 20.

pasture — it moves to find *greener pastures*. And how far does it have to move? Well, it has to move far enough to outrun diffusion. We said before that diffusion wouldn't do any good locally. But suppose it wants to run over there to see whether there is more over there. Then it must outrun diffusion, and how do you do that? Well, you go that magic distance, D/v. So the rule is then, to outswim diffusion you have to go a distance which is equal to or greater than this number we had in our S constant. For typical D and v, you have to go about 30 microns and that's just about what the swimming bacteria were doing. If you don't swim that far, you haven't gone anywhere, because it's only on that scale that you could find a difference in your environment with respect to molecules of diffusion constant D (Fig. 20).

Let's go back and look at one of those sections from Berg's track (Fig. 11). You'll see that there are some little trips, but otherwise you might ask why did it go clear over here and stop. Why did it go back? Well, my suggestion is, and I'd like to put this forward very tentatively, that the reason it does is because it's trying to outrun diffusion.

Otherwise, it might as well sit still, as indeed do the mutants who don't have flagella. Now there is still another thing that I put forward with even more hesitation because I haven't tried this out on Howard yet. When he did his chemotaxis experiments, he found a very interesting behavior. If these things are put in a medium where there is a gradient of something that they like, they gradually work their way upstream. But if you look at how they do it and ask what rules are they using, what the algorithm is to use the current language, for finding your way upstream, it turns out that it's very simple. The algorithm is: if things are getting better, don't stop so soon. If, in other words, you plot, as Berg has done in some of his papers, the distribution of path lengths between runs and the little stops that he calls "twiddles", the distribution of path lengths if they are going *up* the gradient gets longer. That's a very simple rule for working your way to where things are better. If they're going down the gradient though, they don't get shorter. And that seems a little puzzling. Why, if things are getting worse, don't they change sooner? My suggestion is that there is no point in stopping sooner. There is a sort of bedrock length which outruns diffusion and is useful for sampling the medium. Shorter paths would be a ridiculous way to sample. It may be something like that, but as I say, I don't know. The residue of education that I got from this is partly this stuff about simple fluid mechanics, partly the realization that the mechanism of propulsion is really not very important except, of course, for the physiology of that very mysterious motor, which physicists aren't competent even to conjecture about.

I come back for a moment to Osborne Reynolds. That was a very great man. He was a professor of engineering, actually. He was the one who not only invented Reynolds number, but he was also the one who showed what turbulence amounts to and that there is instability in flow, and all that. He is also the one who solved the problem of how you lubricate a bearing, which is a very subtle problem that I recommend to anyone who hasn't looked into it. But, I discovered just recently in reading in his collected works that toward the end of his life, in 1903, he published a very long paper on the details of the *sub-mechanical universe*, and he had a complete theory which involved small particles of diameter 10^{-18} centimeters. It gets very nutty from there on. It's a mechanical model, the particles interact with one another and fill all space. But I thought that, incongruous as it may have seemed to put this kind of stuff in between our studies of the sub-mechanical universe today, I believe that Osborne Reynolds would not have found that incongruous, and I'm quite positive that Viki doesn't.

References

Papers on flagellar rotation referred to in the talk:

Berg, H. C. and Anderson, R. A. *Nature* **245**, 380 (1973).
Silverman, M. and Simon, M. *Nature* **249**, 73 (1974).
Larson, S. H., Reader, R. W., Kort, E. N., Tso, W.-W., and Adler, J. *Nature* **249**, 74 (1974).
Berg, H. C. *Nature* **249**, 77 (1974).

Some recent review articles:

Berg, H. C. *Nature* **254**, 389 (1975).
Berg, H. C. *Ann. Rev. of Biophys. & Biolog.* **4**, 119 (1975).
Adler, J. *Ann. Rev. Biochem.* **44**, 341 (1975).
Berg, H. C. *Scientific American*, August, 1975.

—

5

A Possible New Form of Matter at High Density

T. D. Lee

Columbia University, New York

1. Vacuum

In this talk, I would like to discuss some of my recent theoretical speculations, made in collaboration with Gian Carlo Wick. Over the past year,
we have tried to investigate the structure of the vacuum. It is through this
investigation that the possibilities of vacuum excitation states and abnormal nuclear states have been suggested. Before coming to the main topic,
whether or not there may be the possibility of a new form of matter at high
density, perhaps I should first digress on questions related to the vacuum.

In physics, one defines the vacuum as the lowest energy state of the
system. By definition, it has zero 4-momentum. In most quantum field-
theoretic treatments, quite often the vacuum state is used only to enable us
to perform the mathematical construct of a Hilbert space. From the vacuum state, we build the one-particle state, then the two-particle state, . . . ;
hopefully, the resulting Hilbert space will eventually resemble our universe.
From this approach, different vacuum state means different Hilbert space,
and therefore different universe.

Nevertheless, one may ask: What is this vacuum state? Does it have
complicated structure? If so, can a part of this structure be changed? Ever
since the formulation of relativity, after the downfall of the classical aether
concept, one learns that the vacuum is Lorentz invariant. At least, one knows
that just running around and changing the reference system won't alter the
vacuum. However, Lorentz invariance alone does not insure that the vacuum
is necessarily simple. For example, the vacuum can be as complicated as

the product or sum of any scalar field or other scalar object at the zero 4-momentum limit:

$$\text{vacuum} \sim \phi^n \quad \text{or} \quad (\overline{\psi}\psi)^m \quad \text{at} \quad k_\mu = 0. \tag{1}$$

There is other circumstantial evidence indicating that the vacuum may have complicated and changeable structures. That the vacuum has complicated structure is well known, as shown by the following:

Det Kgl. Danske Videnskabernes Selskab.
Mathematisk-fysiske Meddelelser. **XIV, 6.**

ÜBER DIE ELEKTRODYNAMIK DES
VAKUUMS AUF GRUND DER QUANTEN-
THEORIE DES ELEKTRONS

VON

V. WEISSKOPF

KØBENHAVN
LEVIN & MUNKSGAARD
EJNAR MUNKSGAARD
1936

That this complicated structure of the vacuum may in part be changeable will be discussed on the following page.

1.1. *Missing symmetry*

Symmetry principles have played an important role since the beginning of physics. Apart from their intrinsic aesthetic beauty, symmetry considerations have provided us an extremely powerful and useful tool in our effort to understand nature. Gradually, they have become the backbone of our theoretical formulation of physical laws. Yet, especially over the past two decades, with very few exceptions, most of the symmetries used in physics have been found to be broken. If we consider symmetry quantum numbers such as the isospin $\vec{I}$, the strangeness S, the parity $P, \ldots$, we find

$$\frac{d}{dt} \left\{ \begin{array}{c} \vec{I} \\ S \\ P \\ C \\ CP \\ \vdots \end{array} \right\}_{\text{matter}} \neq 0. \tag{2}$$

Aesthetically, this may appear disturbing. Why should nature abandon perfect symmetry? Physically, this also seems mysterious. What happens to these missing quantum numbers? Where do they go to? Can it be that matter alone does not form a closed system? If we also include the vacuum, then perhaps symmetry may be restored:

$$\frac{d}{dt} \left\{ \begin{array}{c} \vec{I} \\ S \\ P \\ C \\ CP \\ \vdots \end{array} \right\}_{\text{matter+vacuum}} = 0. \tag{3}$$

This is, of course, the basic idea underlying the general heading "Spontaneous Symmetry Breaking". On the other hand, unless we have other links connecting matter with the vacuum, how can we be sure that this idea is right, and not merely a tautology?

A way out of this dilemma is to realize that the restriction $k_\mu = 0$ for the vacuum state is only a mathematical idealization. After all, the universe has a finite radius, and k_μ is never strictly zero. So far as the microscopic system of particle physics is concerned, there is little difference between $k_\mu = 0$ and k_μ nearly 0; the latter corresponds to a state that varies only very slowly over a large space-time extension, say of dimension

$$L \gg \text{(microscopic length)}. \tag{4}$$

The typical microscopic length may be ~ 1 fm for a nuclear system, and smaller for a subnuclear system. Thus, it seems worthwhile to search for solutions in a quantum field theory which vary only over a very large space-time volume.

1.2. *Physics at small distances vs. physics at large distances*

The concept of small and large distances is of course a relative one. In high energy processes of energy-momentum transfers $\sim 10\text{--}10^2$ GeV, the corresponding spatial distance is $\sim 10^{-2}\text{--}10^{-3}$ fm; in such a scale, the typical nuclear length of 1 fm would appear as a large distance. It may be of interest to see how physics changes between these two different scales of distance. For example, we may consider any inclusive processes:

$$A + B \to C + \cdots .$$

Let b be the impact parameter; a measure of b may be the transverse momentum $p_\perp$ of particle C:

$$b \sim \frac{1}{p_\perp}. \tag{5}$$

The dimensionless cross section $(d\sigma/db^2)$ for a typical high energy strong process, say $pp \to \pi + \cdots$ with $b < 10^{-1}$ fm, is found to decrease rapidly as $\approx b^4$ when b decreases. For a typical electromagnetic collision, say between charged leptons, one has $(d\sigma/db^2) \sim O(\alpha^2)$ and relatively insensitive b. For any weak process, say $\nu_\mu + p \to \mu^- + \cdots$, $(d\sigma/db^2)$ is found to increase rapidly as $\sim b^{-4}$ when b decreases.

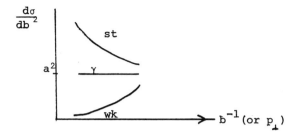

Thus, at present, as the impact parameter becomes smaller and smaller, the difference between strong, electromagnetic and weak interactions diminishes steadily. Any naive extrapolation would lead to the conclusion that at a distance $\ll 10^{-1}$ fm, these different interactions would become comparable in strength.

Suppose this turns out to be indeed the case. At distances, say $\sim 10^{-3}$ fm or smaller, all three interactions do unify into a single one! At such distances, hadrons would then behave like leptons. Nevertheless, there remain some very puzzling problems: why and how at large distances (~ 1 fm) should there exist the observed huge differences between these interactions? Why should there be the enormous dissimilarity between hadrons and leptons? If the basic equations are determined by physics at small distances, then how can the same equations admit solutions which appear to be of a totally different character at large distances?

A possible answer may lie in the direction of string models and bag models for hadrons. Much of the recent work in this area has been done by Professor Weisskopf and his collaborators here at M.I.T. A characteristic of these models is that the observed hadrons are to be described as extensive objects, resembling the macroscopic solutions of either a vibrating string or a resonating cavity. So far as these extensive objects are concerned, the basic law which describes physics at a much smaller distance must appear to be local in character. A deep mystery is then how to connect such "macroscopic" solutions with a local theory. What forms the connecting link between them? Again, there is the possibility that the vacuum may also play an important role here. Perhaps, it is due to the different manifestations of the vacuum that one has these different physical appearances.

With these remarks in mind, we now turn to the question of how to produce the vacuum excitation states.

2. Vacuum Excitation

According to (1), the vacuum is $\sim \phi^n$ or $(\overline{\psi}\psi)^m$ at the zero momentum limit. Its excitation can be viewed simply as a change in the expectation value of the corresponding operator ϕ^n or $(\overline{\psi}\psi)^m$ over a large volume. The simplest phenomenological description is to adopt the language of a scalar local field ϕ. The vacuum state is the lowest energy state in such a local field theory. Through the transformation $\phi \rightarrow \phi +$ constant, one may always assume as a convention

$$\langle \text{vac}|\phi|\text{vac}\rangle = 0. \tag{6}$$

A vacuum excitation is represented by an excited state in which the expectation value

$$\overline{\phi} \equiv \langle \phi(x)\rangle \simeq \text{constant} \neq 0 \qquad \text{inside } \Omega$$
$$= 0 \qquad \text{outside } \Omega \tag{7}$$

where the linear dimension $\Omega^{1/3}$ is $\gg$ microscopic distance.

How can we produce such changes in $\langle \phi \rangle$? If produced, how can we detect them? The problem is analogous to the formation of "domain structures" in a ferromagnet. We may draw the analog:

$$\langle \phi(x)\rangle \longleftrightarrow \text{spin}$$
$$J = \text{matter source} \longleftrightarrow \text{magnetic field}.$$

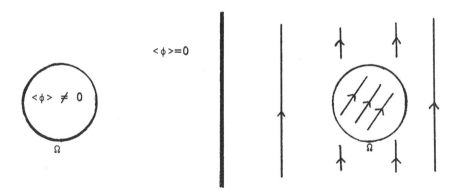

In the case of a very large ferromagnet, because its spin interacts linearly with the magnetic field, a domain structure can be created by applying an external magnetic field over a large volume. Furthermore, after domains are created, one may remove the external field; depending on the long-range forces, the surface energy and other factors, such domain structure may persist even after the external magnetic field is removed. Similarly, by applying over a large volume any matter source J which has a linear interaction with $\phi(x)$, one may hope to create a domain structure in $\langle \phi(x) \rangle$. Depending on the dynamical theory, such domains may also remain as physical realities, even after the matter source J is removed. It is important to note that high temperature tends to destroy domain structures. Therefore, throughout this talk, I will concentrate only on zero temperature or low temperature phenomena.

As an illustration, we may consider a local scalar field theory. The Lagrangian density is

$$\mathcal{L}_\phi = -\frac{1}{2}\left(\frac{\partial\phi}{\partial x_\mu}\right)^2 - U(\phi). \tag{8}$$

If the theory is renormalizable, then U is a fourth order polynomial in ϕ:

$$U(\phi) = \frac{1}{2}a\phi^2 + \frac{1}{3!}b\phi^3 + \frac{1}{4!}c\phi^4. \tag{9}$$

Since we are interested in the long wavelength limit of the field, the scalar field $\phi(x)$ is used only as a phenomenological description. It can be any kind of 0+ resonance between particles. The details of its microscopic structure do not concern us, nor is the renormalizability an important factor. All we are interested in is that its zero 4-momentum limit exists; once such a limit exists, then the expectation value $\langle \phi \rangle$ becomes automatically and inextricably connected to the description of the vacuum and the question of vacuum excitations.

In (9), we require the constant $c > 0$ so that U has a lower bound, and $a > \frac{1}{3}b^2/c > 0$ so that the absolute minimum of U is at $\phi = 0$, in accordance with our convention (6). Since we are only interested in a slowly varying field over a large volume, we may neglect both the surface energy and the energy due to $\left(\frac{\partial\phi}{\partial x_\mu}\right)^2$. Therefore, the classical description of $\phi(x)$ should be a reasonably good approximation. The energy density of the field is then simply $U(\phi)$.

2.1. *Constant source*

Let us now introduce an external source $J(x)$. As a first example, we consider the simplest case that J is a constant inside a large volume Ω, but zero outside. The energy of the system becomes

$$[U(\phi) + J\phi]\Omega.$$

The following graphs illustrate how in the lowest energy state, inside Ω

$$\bar{\phi} = \langle \phi(x) \rangle \tag{10}$$

can be changed under the influence of a constant J:

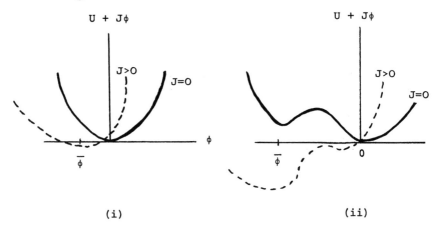

(i)   (ii)

In case (ii), as J increases, there is a critical value at which $\bar{\phi}$ makes a sudden jump. Once a domain of an abnormal value of $\bar{\phi}$ is created, if J is subsequently removed, in case (i), $\bar{\phi}$ will return to 0; but in case (ii), depending on the magnitude of the abnormal value produced, even if J is removed $\bar{\phi}$ may not return immediately to 0.

2.2. Matter source

Let us consider the more realistic case that the source is not a constant, but consists of, say, some physical Fermions, represented by a Dirac field ψ. [Of course, if one wishes, one can also use Bosons instead of Fermions.] For our later discussions on abnormal nuclear states, ψ is the nucleon field; for other purposes such as the "bag model", ψ can be the quark field. The Lagrangian density becomes

$$\mathcal{L} = \mathcal{L}_\phi - \psi^\dagger \gamma_4 \left[\gamma_\mu \frac{\partial}{\partial x_\mu} + (m + g\phi) \right] \psi, \tag{11}$$

where $\mathcal{L}_\phi$ is given by (8), m is the mass of the free Fermion, and g is the coupling constant. An important feature is that if $\overline{\phi}$ is a constant $\neq 0$ over a large volume Ω, then inside Ω the effective mass of the Fermion becomes

$$m_{\text{eff}} = m + g\overline{\phi}. \tag{12}$$

Thus, by measuring m_{eff}, one can detect whether there is a any change in $\overline{\phi}$. In the above Lagrangian, so far as the Fermion field is concerned, there is an equivalence between

$$m \rightarrow m + \delta m$$

and $\tag{13}$

$$\phi \rightarrow \phi - \frac{1}{g}\delta m.$$

Up to a point, this transformation resembles both the gauge invariance in electrodynamics

$$\psi \rightarrow e^{i\theta}\psi,$$

$$A_\mu \rightarrow A_\mu - \frac{1}{e}\frac{\partial\theta}{\partial x_\mu}, \tag{14}$$

and the equivalence between gravitation and acceleration in general relativity. However, in both electromagnetism and gravitation, because of the zero mass of photon and graviton we have exact symmetries. Here, because there is no spin 0 particle of zero mass, transformation (13) is not an exact symmetry in the physical world. [It would, of course, be disastrous if it were, since the inertia mass would then become a non-observable.] As we shall see, this transformation can be used to produce changes in m and ϕ. Thereby, it may lead to a new form of matter at high density.

3. Abnormal Nuclear States

Let the volume Ω be filled with nucleons of density n,

$$n \equiv \Omega^{-1}\int \langle\psi^\dagger\psi\rangle d^3r. \tag{15}$$

As we shall see, if the nucleon density n is sufficiently high, the system may exist in an "abnormal nuclear state", in which the effective nucleon mass becomes zero, or nearly zero (instead of the free nucleon mass m_N). The simplest way to see why such an abnormal state may develop is to examine the quasi-classical solution.

Let us assume that the nucleons form a Fermi gas of a uniform density and ϕ is a classical field. In such a quasi-classical treatment, the lowest energy state is one in which ϕ is a constant and the Fermi gas is completely degenerate. The corresponding energy density $\xi(n)$ is given by

$$\xi(n) = U(\phi) + \frac{2}{\pi^2} \int_0^{k_F} (k^2 + m_{\text{eff}}^2)^{1/2} k^2 dk, \tag{16}$$

where m_{eff} is the effective mass of the nucleon, related to $m_N = 940$ MeV by

$$m_{\text{eff}} = m_N + g\phi. \tag{17}$$

k_F is the Fermi momentum given by

$$k_F = (3\pi^2 n/2)^{1/3}. \tag{18}$$

For simplicity, we assume the nuclear matter to be composed of half protons and half neutrons. We shall also assume Ω to be sufficiently large that the surface energy can be neglected.

Throughout our discussion, we define the *normal nuclear state* to be one in which $m_{\text{eff}} \cong m_N$ and the *abnormal nuclear state* to be one in which $m_{\text{eff}} \cong 0$. As n increases, the Fermi-sea contribution to the energy becomes increasingly more important. Thus, independent of the detailed form of $U(\phi)$, in the high density limit one finds

$$\lim_{n \to \infty} \phi = -(m_N/g) \quad \text{and} \quad \lim_{n \to \infty} m_{\text{eff}} = 0; \tag{19}$$

i.e., the state becomes abnormal. On the other hand, in the low density limit, because of (6), one must have

$$\lim_{n \to 0} \phi = 0 \quad \text{and} \quad \lim_{n \to 0} m_{\text{eff}} = m_N; \tag{20}$$

i.e., the state is normal. As illustrated in the following figure, depending on the parameters in the theory, the transition from the low density "normal" solution ($m_{\text{eff}} \cong m_N$) to the high density "abnormal" solution ($m_{\text{eff}} \cong 0$) may or may not be a continuous one. The mechanism of the transition is similar to that in the case of a constant J discussed before. There is however one important difference: because the nucleon energy depends only on

$(m_N+g\phi)^2$; its minimum occurs at $\phi=-m_N/g$. Hence, $\mathrm{Lim}\,\phi=-m_N/g$ as $n\to\infty$, while in the case of a constant J one has $\mathrm{Lim}\,\phi=-\infty$ as $J\to\infty$.

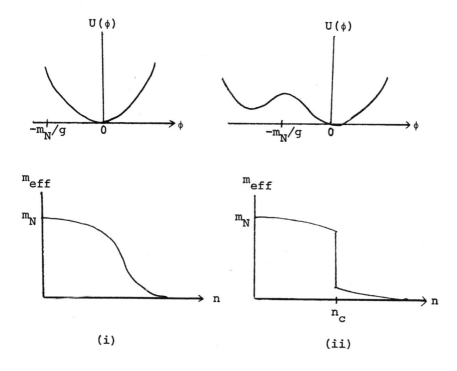

(i)   (ii)

A particularly interesting example of a discontinuous transition is the well-known σ-model. For a fairly wide range of the parameters, the critical density n_C for the transition is found to be approximately given by

$$n_C \cong 11.6\left(\frac{m_\sigma}{m_N}\right)^2\left(\frac{g^2}{4\pi}\right)^{-1} n_0 \qquad (21)$$

where

$$n_0^{-1} = \frac{4\pi}{3}(1.2\ \text{fm})^3,$$

and m_σ is the σ-meson mass. In the following figure, we map out the region in $g^2/4\pi$ and m_σ for $n_C \lesssim 2n_0$.

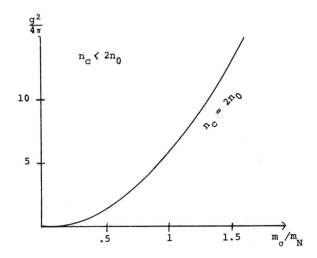

At present, there is no reliable data on either m_σ or g^2. If we identify σ to be the broad 2π resonance, then

$$m_\sigma \sim 700 \text{ MeV}. \tag{22}$$

Extensive calculations have been made in the literature to fit the central attractive part of the nuclear forces from such a 2π exchange. Depending on the assumption of repulsive forces (due to vector meson exchanges), different authors obtained different coupling constants for the 0+ channel:

$$\frac{g^2}{4\pi} \quad \text{varies from 2.5 to 15.} \tag{23}$$

Even with such a wide range of estimated values, it seems possible that the physical reality may be located at the left of the curve $n_c = 2n_0$. Thus, by doubling the present nucleon density one may produce the abnormal nuclear state.

4. Production and Detection

In order to produce the abnormal nuclear state, we must consider reactions in which (i) a large number of nucleons are involved, so that the surface energy can be neglected, and (ii) the nucleon density can be increased by a significant factor ~ 2, One is, therefore, led to considerations of high energy collisions between heavy ions, say

$$U + U \rightarrow \text{Ab} + \cdots \qquad (24)$$

where Ab denotes the abnormal nuclear state. If Ab is produced, what are its characteristics? How can it be detected?

Because of the high density involved, the equation of state of the abnormal state depends sensitively on the short-range repulsive force. For definiteness, we may assume it to be a simple hard-sphere repulsion. Let d be the diameter of the hard spheres. In the following figure, we plot the (volume) binding energy per nucleon,

$$\text{b.e.} \equiv m_N - (\xi/n),$$

versus

$$d \equiv \text{hard-sphere diameter:}$$

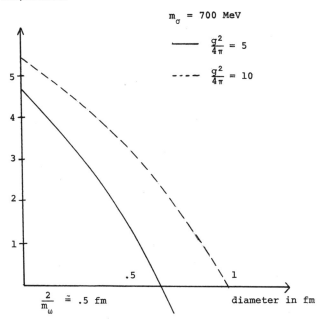

As an example, we may assume $m_\sigma = 700\,\text{MeV}$, $(4\pi)^{-1}g^2 = 5$ and $d = 2/m_\omega \cong .5\,\text{fm}$ where $m_\omega = \omega^0$ meson mass. In this case, per nucleon the kinetic energy $\sim (940 - 490 - 300) \sim 150\,\text{MeV/nucleon}$ for the abnormal state. The nuclear radius is $\sim (.9\,\text{fm})A^{13}$. If one neglects both the surface energy, and Coulomb energy, the abnormal nuclear state is *stable*! [For a pure neutron system, the kinetic energy is much higher because of the exclusion principle; in addition, due to ρ^0-exchange, one expects the repulsive force between neutrons to be greater than that between neutron and proton. Consequently, most likely, the abnormal state of a pure neutron system is unstable.]

A clear signal of reaction (24) is the detection of a stable (or metastable) nucleus of very large baryon number A, say, ~ 400. For a charge $Z \sim 200$, the Coulomb energy is not important; therefore, we expect $Z \sim \frac{1}{2}A$. Once an abnormal state is produced, through successive neutron absorption one may hope to increase its A and Z gradually. When Z increases, Coulomb energy becomes important; furthermore, the abnormal state can create e^+e^- pairs. The e^+ will be sent to infinity, but a fair fraction of the e^- will be kept within the abnormal nucleus. As Z increases, the number of e^- also increases. The interplay between the added Fermi energy of e^- and the Coulomb energy may eventually bring the abnormal state to the point of instability when A reaches $\approx 10^4$.

There remains the problem of the production mechanism of the abnormal state. Only a brief discussion will be given, since we are not able to evaluate the various complexities involved. Let us consider reaction (24) at $\sim \frac{1}{2}$–1 GeV/nucleon in the center-of-mass system.

(i) Penetration problem:

Although the mean free path of a nucleon inside the nuclear matter is relatively short, because of the high energy, in most of the nucleon–nucleon collisions one expects the two final nucleons to move more or less along their initial general direction, with about 1 or 2 soft mesons created per nucleon–nucleon collision. Thus, provided the impact parameter is small, these two U nuclei should penetrate each other, and thereby increase the nuclear density from its normal value n_0 to $\sim 2n_0$. In addition, there is a large number of soft mesons produced which carry away a large part of the excess energy.

(ii) Response time:

The collision time of (24) is about

$$\tau_{coll} \sim 10^{-12} \, cm/c$$

where c is the velocity of light (provided that the energy is not very high so that Lorentz contraction can be neglected). The order of magnitude of the response time of the meson field is about

$$\tau_{resp} \sim \frac{h}{m_\phi c} \qquad c \sim 3 \times 10^{-14} \, cm/c.$$

This follows from the simple Klein–Gordon equation for a scalar field ϕ:

$$-\frac{\partial^2}{\partial x_\mu^2}\phi + m_\phi^2 \phi = -g \, n_S + \cdots$$

where $\cdots$ denotes the non-linear part of the meson–meson force and $n_S = \langle \psi^\dagger \gamma_4 \psi \rangle$. Since

$$\tau_{resp} \ll \tau_{coll},$$

as n_S changes, the value of the meson field responds almost adiabatically. when n_S reaches $\sim m_\phi^2 m_N/g$, the meson field becomes $\phi \sim -m_N/g$ which forms the abnormal state.

(iii) Potential barrier:

After the two U nuclei penetrate each other, and ϕ develops the abnormal value, there is still the question: would the nucleons keep on moving away from each other, so that the whole system becomes separated? We note that as the meson field assumes the abnormal value $\phi \sim -(m_N/g)$, the effective nucleon mass becomes zero. The energy of each nucleon is $|\vec{p}|$ inside the nucleus, but $(\vec{p}^2 + m_N^2)^{1/2}$ outside. Thus, there is a potential barrier B holding the nucleons together in the abnormal state. So long as $|\vec{p}|$ is $\leq m_N$, most of the nucleons will collide with the barrier and remain inside. Through such collisions, additional mesons will be emitted, and the nucleons may gradually readjust their wave functions to those of the abnormal state.

While an accurate calculation of the production probability is difficult, from the above discussion, one sees that in order to produce Ab, the collision energy of the heavy ions should not be small; otherwise penetration becomes difficult. The collision energy should also be not too high; otherwise the collision time τ_{coll} may become too small because of Lorentz contraction, and the barrier B becomes $\sim \frac{1}{2}|\vec{p}| - 1m_p^2$ which may also be too low to hold the nucleons together. The best value may perhaps be ≈ 100 to several hundred MeV/nucleon in the C.M. system.

5. Remarks

The question whether we live in a "medium" or in a "vacuum" dates back to the beginning of physics. From relativity, we know that the "vacuum" must be Lorentz-invariant. As remarked before, Lorentz invariance by itself does not mean that the "vacuum" is simple. From Dirac's hole theory, one has learned that the vacuum, though Lorentz-invariant, can be rather compli-cated. However, so long as all of its properties cannot be changed, so long as, e.g., the value of vacuum polarization cannot be modified, then it is purely a question of semantics whether the vacuum should be called a medium or not.

What we try to suggest is that if we do indeed live in a medium, then there should be ways through which we may change the properties of that medium.

Hitherto, in high energy physics we have concentrated on experiments in which we distribute a higher and higher amount of energy into a region with smaller and smaller dimensions. In order to study the question of "vacuum", we must turn to a different direction; we should investigate some "bulk" phenomena by distributing high energy over a relatively large volume. *The fact that this direction has never been explored should*, by itself, *serve as an incentive for doing such experiments.* As we have discussed, there are possibilities that abnormal states may be created, in which the nucleon mass may be very different from its normal value. It is conceivable that inside the volume of the abnormal state, some of the symmetry properties may become changed, or even that the usual roles of strong and weak interactions may become altered. If indeed the properties of the "vacuum" can be transformed, we may eventually be led to some even more striking consequences than those that have been discussed in this lecture.

6

The World As Quarks, Leptons and Bosons*

Murray Gell-Mann
California Institute of Technology

I hope that my few disconnected remarks will not appear too inadequate after that magnificently organized presentation by T. D. Lee. I remember very well the time, more years ago than I would like to count, when I faced the choice between graduate school and M.I.T. and suicide. I chose to come here, actually, after a difficult decision. That it turned out to be the right decision became clear when I met the man who had hired me as his assistant, the man that we are gathered here to honor today — a man known all over the world by his nickname, "Viki". I had better be careful, especially after the previous speech, when I do get to discussing elementary particles, not to dwell too long on symmetry breaking by the vacuum. I understand that the press is represented here and I can visualize in the Boston newspapers the headline "M.I.T. Physicist Honored by Speeches About Nothing".

It was a profoundly important experience for me to meet Viki Weisskopf at that time. I learned a great deal from him, not only about the atomic nucleus but also all sorts of lessons, especially about avoiding cant and pomposity — something very important in life in general, but particularly in theoretical physics; and prizing agreement with the evidence above mathematical sophistication; and always to search, if at all possible, for simplicity. We learned, all of us. I notice that a number of my fellow students are here

---

*Editor's Note: This is a transcription from the tape of Professor Gell-Mann's talk, and he has not checked it for accuracy. The reader should also be aware that this talk was delivered before the existence of the J-particle was known to other than its discoverers.

today, including Larry Biedenharn, who was my officemate and roommate. We learned in informal discussions at the blackboard, in Viki's office, in each other's offices, and in discussions with the post-docs who were here at the time. I see several of them here now — Bruce French, Larry Spruch, and Murph Goldberger. We all talked together at the blackboard, screaming and yelling at one another, arguing about different theories, about the interpretation of experimental evidence. Here at M.I.T. all calculations were done in Weisskopf units in which $4\pi = 1, 2 = 1, -1 = 1$. Occasionally "i" was put equal to 1, which sometimes gave serious difficulties in the interpretation of the results.

We learned a lot about many things too in evening sessions, in dinner sessions to which Viki gave an enormous amount of his time. We discussed politics, philosophy, physics — anything as long as it was not pretentious. We also learned things in class. I understand that in the introductory lecture of this Symposium, Viki Weisskopf was described as one of the leading teachers and one of the greatest teachers at M.I.T.. That's certainly true, although I'm not sure everybody knows exactly in what sense that's true. For example, we had a great many polished lecturers at that time, both here and at Harvard, and I remember attending their classes. From some of them I learned things, but usually by the next day I had forgotten whatever it was I had learned. That was never true in the case of Viki Weisskopf's classes. For example, he once tried to derive the resonance formula in a few lines. You know that the first discussion of resonances in quantum physics was in his dissertation, and in the paper by Weisskopf and Wigner. He was explaining the formula to us, in particular as it was applied to nuclei. He had been the leading exponent of nuclear resonances for several years preceding that time, and was at work on a book including many chapters on them. The first derivation on the blackboard came out with the Weisskopf–Wigner formula upside down. The second derivation had the denominator in the right place and the numerator in the right place but the width had the wrong sign. He promised that the next time he would have a very simple and elegant derivation. And the next time, indeed, the derivation was correct and very quick, but, by that time, every student had gone home and learned how to do it himself. It is a fact that his manner of teaching was far more effective than that of all the famous polished lecturers that I had ever heard of, and I have never encountered a better teacher than Viki.

Let me say in the few minutes that are left a few words about physics. You saw in T. D. Lee's transparency the cover of a very important paper by Viki about quantum electrodynamics in the 1930's. Quantum electrodynamics is a very significant theory. It embodies, as a field theory, the most important principles in which we believe in physic today, namely, relativity, microscopic causality, and quantum mechanics. These lead to field theory. And if we make a field theory of electrons and electromagnetism in the form of photons, we have quantum electrodynamics — a highly successful theory, the only really successful theory we had. It works to n decimal places at low energies where it's been investigated exhaustively. No mistakes have appeared in it in the 45 years since its construction, It had to be reinterpreted a little bit with infinities being messed around with and so on, but basically it's the same theory that was invented 45 years ago on the basis of the development of quantum mechanics 50 years ago. It's been investigated at high energies up to several GeV's, and it still works, although there the investigations are not to such high accuracy as they are at low energy. We expect eventually some sort of correction will have to be made to it, perhaps at 50 or 100 GeV, but we hope that that correction may be in the form of simply enlarging the theory to include the other phenomena of physics. That sort of synthesis is the thing that elementary particle physics strives for. Our field is not one which is widely known for its immediate applications, although there are some; there are even medical applications. But it's not primarily in search of immediate applications that we study particle physics, but rather in the hope of achieving a grand synthesis at the sub-nuclear level, — the description of all the basic particles, or as many of them as we can encompass in a grand synthesis —, similar to the synthesis of 50 years ago, when with the discovery of quantum mechanics the atom was essentially completely explained except that the nucleus had to be treated as a point. Dirac, when he formulated his relativistic theory of the electron, is said to have remarked that it explained most of physics and the whole of chemistry. We are looking for that sort of synthesis at the sub-nuclear level today. It may be that we are not terribly far from it; that has not really appeared to be the case until recently; but now one can say that perhaps there is a chance we are not very far from describing this realm in such a way. In the history of science such syntheses have never been complete. There have always been a few phenomena, perhaps obscure at the time, which escape description in the grand synthesis. They are sort of the tip of an iceberg, the iceberg itself being the subject matter of the next grand synthesis, if people have the patience to go on to one. Maybe such things will be true again if we do succeed.

The interactions that we know of include gravity, which is extremely weak on the level of particles because they have such small masses. Presumably gravity is described either exactly or almost exactly by Einstein's theory of so-called general relativity, (which is really Einstein's theory of gravity), and is mediated by a quantum graviton with deducible properties. Nobody has the vaguest notion how to find this graviton experimentally, — which makes it rather safe from the theoretical point of view. The other interactions we know of are electromagnetism, mediated by the photon, described beautifully by quantum electrodynamics; the weak interaction, presumably mediated by an intermediate boson $X^{\pm}$, being sought; and other associated interactions just now being found now. Then there is the strong interaction responsible for the nuclear force. Perhaps the CP violation or the T violation discovered 10 years ago may require another interaction. The hope is to try to incorporate these interactions into some sort of unified scheme, and that has appeared in the past to be a very distant hope, but it appears much closer now, — perhaps an illusion.

The strength parameter for electromagnetism is a $\alpha \simeq \frac{1}{137}$. For the strong interaction at typical energies of a GeV or so, it seems to be somewhat bigger although not necessarily gigantic, maybe $\frac{1}{5} - \frac{1}{3}$. The definition of this is somewhat in a state of flux. We used to think a long time ago we knew how to define it, and it was very large. It seems now it is rather small at typical energies. Also the definition may vary with energy, so that the effective coupling constant would become even smaller at very small distances and perhaps much larger at much larger distances. The weak interaction you know about is responsible for β-decay and all sorts of related phenomena. For the weak interaction, as you all know, we don't measure the strength parameter directly, but instead we measure the Fermi constant $G \propto e'^2 / M_X^2$ where e' is a dimensionless parameter and M_X is the mass of the hypothetical intermediate boson. I call it hypothetical, but it's very hard to see how we are going to survive if it isn't found at some reasonable mass. If you assume, as people do today, that there is a sort of unity of the interactions, then you set the strength parameter for the weak interactions roughly equal to the one for electromagnetism, and you get a mass for the intermediate boson of something like 50 to 100 GeV. That is (fortunately) just above where we can detect it with the existing machines, but we hope that either with a stretching of the existing machines or the construction of a new one we really will be able to find this elusive particle. As I say, the quantum mechanics of the weak interaction can hardly survive without it. The experimental lower limit on its mass is only 5 to 10 GeV but that's already rather high.

You can see the immense similarity in the properties of the photon and this intermediate boson X: The photon is electrically neutral, the X is charged. The photon moves with the speed of light, has rest mass zero, the X has rest mass of 50 GeV or 100 GeV — nearly zero, but there's an appreciable difference. For the photon the coupling is parity invariant, or left–right symmetrical. For the X we know that it is purely left-handed. Apart from these minor differences the couplings are virtually identical, and present a very close mathematical analogy which any field theorist can perceive at once. It seems a very satisfactory theory of weak and electromagnetic interactions can be constructed if and only if you unify them, together with some other new associated interactions, including one (at least) that would have to be carried by another intermediate heavy boson Z^0. So you have these quanta, photon, $X^{\pm}$, Z^0, perhaps some others, in a so-called unified gauge theory or generalized Yang–Mills theory. A number of people have worked on this. People with the courage to publish ideas in this field have included Glashow, Salam and Ward, Weinberg, and a number of others. A particular mechanism of making soft mass, invented by a great many people, is often called the Higgs mechanism, after a man who was perhaps the first to do it. Others say we should call it Higumanak or something, after the initials of the many people who worked on it.

You can convince yourself rather quickly that this Yang–Mills theory or gauge theory is really the only sensible basis for making such a unified theory, because trying to make a field theory behave sensibly at high energies, — not in perturbation theory, but by summing all orders of perturbation theory — has proved over 25 years of disappointing efforts to be very difficult, probably impossible. And one is reduced to trying to make the theory behave sensibly, or at least nearly sensibly, in each order of expansion in the coupling constant. But if you do that, then you need to make it a so-called renormalizable theory, — one that doesn't, except for trivial infinities, involve any infinite quantities in the expansion in perturbation theory. Quantum electrodynamics is like that, and the problem was to make the weak interaction theory like that. The reason hideous divergences of quadratic and quartic character have been encountered in previous attempts to make weak interaction theories is very simple. You can trace it, if you look at the diagrams in the field theory, to the fact that the weak charge operator, analogous to the electric charge for the photon, would come along in the diagram and then stop. It was not conserved. Likewise the hermitian conjugate of the weak charge operator would go along some line and then stop. The weak charge was not transferred to another particle or maintained by the particle that was carrying it. This stopping, this explicit non-conservation of the weak charge, even at high frequencies, was the source of all these very bad divergences. The only way to get rid of them is to make a theory in which, at least at high frequencies, the weak charge as well as the electromagnetic charge is conserved. (At low frequencies the non-conservation of the weak charge can be allowed.) But the only mathematical theory of non-commuting conserved charges is just such a Yang–Mills gauge theory, and so a theory something like that had better be at the root of the unification. It had better be at the root of the correct description of the weak interactions.

What sort of particles have these interactions? What we have are lots of leptons: electron and positron, neutrino and anti-neutrino, and then for God knows what reason, another set — muon, anti-muon and its neutrino and anti-neutrino. Rabbi, you remember, asked about those things — "who ordered them?" And we still don"t know in fact why we need this doubling.

Professor Telegdi, who I see is here in the audience, once tried to impress Paul Adrian Maurice Dirac by telling him that the electron was not the only Dirac particle that was known, that there was another called the "muon". But Dirac didn't look impressed. He just said, "Oh, really?" Later, in conversations with me, he referred to it as "Telegdi's new particle, the muon."

Anyway, these leptons feel electromagnetic and weak interactions, and presumably this new neutral associated interaction; or more, if there are more. But they do not feel a strong interaction. When they go into nuclei, they don't notice that the strong interactions are taking place inside the nucleus. They go right through, or at best, if they are charged, they feel the electrical force.

The strongly interacting particles are the baryons, mesons, anti-baryons, and so on, like neutron, proton, pion, and so on. In textbooks, which are always years and years behind, they may try to tell you that these are elementary particles; but that's clearly absurd. These are the particles, though, that you see in the laboratory, that feel the nuclear force, and they have an enormous spectrum. The neutron and proton are stable, or meta-stable, only because they have the lowest energy levels, but there is an enormous, presumably infinite spectrum of baryons of which the neutron and proton are just the lowest state. The mesons also have a huge, presumably infinite spectrum, of which the pions are the lowest state. These spectra include all angular momenta, both parities, anything you want. Nobody in his right mind today would suppose that the neutron and proton are any more elementary than any of the others. Hundreds of states of these things are known, baryons, mesons, and so on, and the elucidation of the spectrum reveals some very interesting simplicities. These particles come in families. The spectra show an enormous amount of symmetry, somewhat broken symmetry, but broken in especially beautiful and understandable ways. The states are arranged in families and super-families and super-super-families, with very nice principles giving the splittings among them. The neutron and proton belong to such a large super-family, so do the pions; and it's crazy to suppose that the neutron and proton are elementary and that their sisters and cousins are not.

So if we are to look for fundamental entities, we must look for them either by making all of the hadrons elementary as the "dual" people do, or none of them elementary, as the now defunct bootstrap approach did. It's been replaced by the dual approach. The dual approach probably works, and it probably will be equivalent to the kind of thing I will talk about, namely, the description of hadrons in terms of quarks and gluons. I have nothing against the approaches that treat the particles without constituents inside. I assume they will end up giving theories, correct and equivalent to the quark–gluon theory. But the most directly successful approach has been the one trying to interpret the hadrons in terms of hidden constituents which are elementary.

Now you must remember that the hadrons, as well as the leptons, are affected by the electromagnetic, weak and associated interactions — these are common to both sets. In fact, a new experiment done at CERN and at NAL with anti-neutrino's hitting nucleons has very likely revealed the first of these new associated neutral interactions, the Z^0 interaction. CERN people had probably seen it some years ago, but it's only recently that they put forward their results. Confirmatory experiments at NAL seem to make it perfectly clear that this interaction does exist, although the Argonne National Laboratory has done some other experiments, which, if they are to be believed, may cast doubt on the identification of this thing with Z^0. But they are very preliminary, fortunately, and so we theorists are not very upset yet. (It takes a little more than that to upset us.)

The picture I want to talk about is one of constructing the hadrons (baryons, mesons and anti-baryons), out of hidden but elementary constituents called "quarks", and also gluons, which are the things that would hold them together. The gluons would be analogous to the photons and $X^\pm$ and so on, that we talked about in connection with the weak and electromagnetic interactions. They are the quanta of the strong interaction holding the quarks together to make mesons and baryons. This quark picture gives a fairly reasonable description of the meson and baryon spectra — the families and super-families; the symmetries of the strong couplings (the vertices of mesons and baryons) as deduced from experiment; and especially the symmetry properties, and also more detailed properties, of the electromagnetic and weak couplings of baryons and mesons. Soon we'll also have the details of the new neutral current interaction for baryons and mesons and be able to compare that. The deep inelastic lepton–baryon reactions give more details about these electromagnetic and weak couplings, more than just their symmetry properties, but something about the dynamical details of the interaction.

You remember what the quarks are. The derivation of the word is obvious, of course. We needn't dwell on that. The quarks are these hypothetical, fundamental sub-units of the strongly interacting particles. If we plot the "mechanical mass" (since these things probably don't really exist, they won't have an actual mass) against the charge ($\frac{2}{3}$, $-\frac{1}{3}$), there's an isotopic doublet u (up) and d (down), and an isosinglet s. They all have spin $\frac{1}{2}$ parity $+$, where $+$ is the parity of the proton. The anti-quarks have spin parity $\frac{1}{2}^-$, because they have spin $\frac{1}{2}$ and obey Dirac's equation. Dirac, by the way, attended some lectures that I gave on quarks at Cambridge about 1966. He came to every lecture — I was very surprised to see him there. He listened in his usual way; that is, he fell asleep at the beginning of each lecture, woke up at the end, and asked questions that showed that he had understood the entire talk from beginning to end. And he indicated that he was very enthusiastic about this quark picture. It's now very popular; most of my colleagues, I think, accept this sort of thing. But at that time, it was not especially popular and I was very surprised that Dirac seemed to like it. I asked him why he was so fond of the quark picture and he said, "Well, they do have spin one half, you know."

They come in "colors", and, if we are patriotic in the United States, Britain or France we can use the colors of the tricolor — red, white, and blue for these. Color, of course, is not to be taken in the real sense. You see, each quark seems to come in three varieties, but this variety, or color or whatever, is a hidden variable. In Viki's native country, I tried to label these things with the color of the flag but I fell on my face with rot, und weiss, und rot. The Austrians have equipped themselves with a degenerate flag that doesn't lend itself to this sort of description. We should, as scientists, actually have ignored all these political considerations, and taken the primary colors, so that we could refer to those particles which are averaged over color and which we actually see as "white", — but we didn't do that. Anyhow, we supposed that in what I'll call the standard quark picture — I doubt that I'll have time to get to any of the heresies today — only colorless states, only the ones where the color is all averaged over and smeared out, are actually seen as hadrons. So quarks, which are colorful, are concealed, and never observed experimentally. There are variant pictures in which this is not true, but the typical picture is one in which we will never see the quarks in isolation. The gluons probably are colorful objects too, color octets instead of color triplets, and we will probably never see them in isolation either. The theory is nevertheless a sensible, falsifiable theory — one that you can verify or prove false by experiment — since indirectly, by observing the hadrons, you can see whether they have as if they were made up of bound quarks and gluons, permanently bound inside. The idea is that colorful objects are permanently bound inside and everything you see, neutron, proton, pion and so on, is an object in which the color has been averaged over and washed out. Presumably this mechanism is associated with the gluons. The gluons, being colorful themselves, are binding themselves in by a non-linear gluon–gluon–gluon interaction. Theories of that kind are not necessarily impossible to construct. In fact, the most elegant theory might have that property.

The observed meson states, the so-called non-exotic ones — all the resonances that we see —, all have principally the configuration $q\bar{q}$ and in particular, the one in which color is washed out:

$$q(\text{red})\ \bar{q}(\text{red}) + q(\text{white})\ \bar{q}(\text{white}) + q(\text{blue})\ \bar{q}(\text{blue}).$$

Technically, this is called a color SU(3) singlet. There are also little bits of $q\bar{q}q\bar{q}$ and a certain amount of glue. But over most of the range of the critical parameter, ξ, (which I probably don't have time to define), you see just $q\bar{q}$. Exotic meson states are also possible, which mostly would have in simplest configuration $q\bar{q}q\bar{q}$, and then go higher. These exotic ones have never been identified, but they could exist. They are certainly there in the continuum. The question is only a dynamical one of whether they exist as resonances. Furthermore, there may also be quarkless mesons which are just balls of glue. You would never see those pure. They would have just glue, and then $q\bar{q}$ (a little bit), then glue and $q\bar{q}q\bar{q}$, and so on. The way you would see them is the following; around 1000 MeV, according to the quark model, nine scalar mesons are supposed to be seen; and no doubt they are there because nearly nine have been seen. But maybe there are 10. Professor Lee, for example, showed this vague bump that people claim they see at 700 MeV or so. That might be a tenth one. It's possible that's not included in the nine $q\bar{q}$ scalars around 1000 MeV. If it's a tenth one then it would appear alone, because the other nine are already accounted for by $q\bar{q}$. It would probably be one of these quarkless ones, and would be a sort of vacuum excitation, and would represent at least something in common between what I'm saying and what he would say. Anyway, I think it's a very important question whether these quarkless mesons exist experimentally.

The baryons are, basically, three quarks qqq, one of each color, red, white, and blue. In particular, the totally anti-symmetric configuration,

$$q(\text{red})\, q(\text{white})\, q(\text{blue}) + q(\text{blue})\, q(\text{red})\, q(\text{white})$$
$$+ q(\text{white})\, q(\text{blue})\, q(\text{red}) - q(\text{white})\, q(\text{red})\, q(\text{blue})$$
$$- q(\text{red})\, q(\text{blue})\, q(\text{white}) - q(\text{blue})\, q(\text{white})\, q(\text{blue})$$

seems to be the only possible one and that's the one in which the color is washed out — the color SU(3) singlet, to mathematicians. That's what we identify with the real baryons, like the neutron and proton. And, of course, they also have little bits of $q\bar{q}$ and so on, but not much. Again, you could have exotic states starting with three quarks and a $q\bar{q}$, and going higher. None of these is known except perhaps one. If that one is there, it must be accompanied by dozens of others according to this picture. So it's a very important question to see whether that one is really there, or whether we can talk the experimentalists out of it. Either that, or talk them into finding a lot more, but that's an important question. I say all these things because I would like to revive this dying science of bump hunting. Many physicists seem to be ashamed to be engaged in the spectroscopy of baryons and mesons. Unlike atomic spectroscopy, this is not a field in which the fundamental equation was supposed to have been written down 50 years ago, it's still very important to find out about the spectroscopy of hadrons. I think the work should be pushed with great vigor.

I won't bother to go over the agreement with experiment — I think most of you have heard about it. In fact, the mesons do look like $q\bar{q}$. The lowest states, singlet S and triplet S, nonet and a vector nonet, work perfectly. For baryons, the lowest state would be an octet and a decimet. They have been anti-symmetrized in color, so they get symmetrized in everything else. If the spatial configuration is symmetric, which would be natural for a ground state, then they would be symmetrical in spin and in the u, s, d variables. There are six states, then, for the quarks: spin-up, spin-down, and u, s, d make six possible states for each quark. If we totally symmetrize, then we have $\binom{8}{3} = 56$ states. Now if you take the quartet in the decimet configuration (40 states) and the spin doublet in the octet configuration (16 states) you get 56 states, which is exactly what you see. It's not surprising that the color has worked out to give this result, because that's why it was invented. However, as we'll see, there is another very striking independent confirmation of the color idea. We'll get to that in a moment.

If we go on to the next levels, they work too, more or less. Here you get P waves of the q and $\bar{q}$: Tensor mesons, axial vector mesons, the nine scalar mesons that I spoke of, and nine of another kind of axial vector. For each of these, some have been seen: all nine in the case of the 2^+, most of them in the case of the 0^+, and a couple in the case of 1^+. The nonets remain to be filled out. Presumably, that's the most believable part of the thing, anyway, that you get nonets. Still, the experimental work must continue until we really see that this is the pattern of the next states with the opposite parity and not something different — with an intrusive state that we hadn't expected, or something of that kind; or, as we mentioned a moment ago, a tenth one for the scalars (the quarkless one), which would not contradict the hypothesis, but would be very interesting, because we don't know whether it should be there or not. For the baryon, it's the same thing roughly. That is, you look at the next excitation with reverse parity and there should be a so-called 70 ($l = 1$) — 70 states grouped into singlets, octets, and decimets. Again, everything seems to be there, but some of the multiplets are not yet filled out experimentally. It would be a good idea to do so, to make sure that there isn't anything floating around that isn't in the list. So far nothing has been found that isn't in the list in this energy region, but we must make sure. More experimental work and analysis is needed. It's not just experimental physics here but huge amounts of computer phase shift analysis and other boring stuff which, however, is very interesting in its final results, namely, the checking of the spectrum.

The most important thing about the quarks is that their electromagnetic, weak, and presumably the new associated neutral interactions, are supposed to be just about the same as those of leptons, — a genuinely universal coupling —, so that the quarks appear just as elementary as the electron, muon, neutrino. These are the formulae for the interactions:

$$-[i\bar{e}_L \gamma_\alpha e_L + i\bar{e}_R \gamma_\alpha e_R]A^\alpha.$$

Here's the old quantum electrodynamics formula. Dirac's gamma matrix between electron and positron. Here's the electromagnetic potential. The muon looks exactly the same way. Quark electromagnetic current:

$$J_\alpha^{em} = \frac{2}{3}i\bar{u}_L \gamma_\alpha u_L + \frac{2}{3}i\bar{u}_R \gamma_\alpha u_R$$

$$- \frac{1}{3}i\bar{d}_L \gamma_\alpha d_L - \frac{1}{3}i\bar{d}_R \gamma_\alpha d_R$$

$$- \frac{1}{3}i\bar{s}_L \gamma_\alpha s_L - \frac{1}{3}i\bar{s}_R \gamma_\alpha s_R$$

(summed over the three colors).

The quarks have exactly the same interaction except they have different charges: Instead of 1,

$$+\frac{2}{3}, \quad -\frac{1}{3}, \quad \text{and} \quad -\frac{1}{3}.$$

The coupling of quarks to the weak interaction is supposed to be exactly parallel to that of leptons:

$$(\bar{e}_L \gamma_\alpha v_e) X^\alpha.$$

The electron and neutrino, coupled by Dirac's same matrix. This is why we say mathematically there is a striking parallel between the electromagnetic and the weak interaction. The coupling is really the same. The only difference is the projection on the left-handed electron, i.e., the electron interacts only in its left-handed state for the weak interaction, and it interacts equally in left and right-handed states for the photon. The muon and its neutrino have exactly the same interaction. Quark weak current:

$$J_\alpha^{\text{weak}} = u_L \gamma_\alpha (d_L \cos \theta + s_L \sin \theta)$$

(summed over the three colors).

Here you have the up-quark u and a linear combination of down and singlet quarks, d and s, with a mysterious angle θ which is experimentally around 15 degrees. Nobody has the vaguest idea why.

Recently, in order to make these unified theories work, Glashow, Illiopoulos, Maiani and perhaps various other people suggested (I am sorry I paid less attention to it than I should have in previous years) what sounded at the time like a strange idea, that there should be perhaps another dimension for the quarks. There should be another kind of quark, called charmed quark, as opposed to strange quark. It would be coupled to the other linear combinations: $s' = s \cos \theta - d \sin \theta$. If you do that it makes it a lot easier to reconcile a unified theory with experiment. So many people are going to look for such charmed quarks. Since you can't see the quarks anyway, how do you look for a new charmed quark? You look for charmed hadrons. The expectation is that at a couple of GeV excitation above ordinary neutrons and protons, one should see the phenomenon of strangeness discovery repeated again — the phenomenon of charm. I believe it was Karl Marx who said that historical events happen twice; once seriously and the second time as farce. And if the charm is discovered, this would be an example of that. In the case of the strangeness, the decay Q value was about 150 MeV typically. That's the separation between a particle of one strangeness and particle of another strangeness. The decay lifetimes were $\sim 10^{-10}$ sec. and you could see the strange particles in a cloud chamber, a bubble chamber, a track chamber; the neutral ones as gaps, and the charged ones as tracks. But these charmed particles, if they exist, would decay more rapidly, because they are more highly excited (like a couple of GeV) and so there would be very little gap or very little track. You would have to detect them by their propensity to decay into leptons plus hadrons, because they would decay by weak interactions. So you would have to use the leptonic decays of highly excited hadrons as a signature and look for the associated production of charmed and anti-charmed particles. You can do this with strong interactions or you can do it with electromagnetic ones. Such experiments are under way and so far the results are inconclusive, but maybe somebody will find it. If so, then we simply have to add this charmed quark (which also has charge $+\frac{2}{3}$) to the list. Otherwise the quark hypothesis remains unchanged.

The quark picture is in beautiful agreement with the various selection rules for electromagnetic and weak interactions. All the symmetry properties work. I give you an example of the most striking one. The Σ^- particle does decay as predicted into $ne^-\bar{v}_e$. You might think the Σ^+ would then like to decay into ne^+v_e, but a selection rule that comes from the quark model doesn't allow it. Many thousands of these decays have been looked at in the case of the Σ^- and so far there is no corresponding decay of the Σ^+. So to lowest order in the weak interaction, the selection rules that come from the quark model.

Recently, new experiments of a different kind have explored the dynamical features of this coupling more closely — deep inelastic experiments in which in electron or a neutrino is fired at a nucleon, high energies and very high momentum is transmitted to the outgoing lepton. In the limit, as the momentum transfer gets large and the energy gets large and the ratio of these is kept constant, you observe all kinds of very interesting phenomena which were studied theoretically by Bjorken. You sum up over everything that's produced: So, electron plus proton gives electron plus anything, and you add up all the anythings. There's the Bjorken parameter ξ, which I will not bother to define because we are in a hurry. It goes from 0 to 1. For $\xi = \frac{1}{4}-1$, the nucleon appears in these experiments to be made essentially of three quarks. You expect theoretically and you find experimentally, that for a small value of ξ there are also a lot of pairs and other stuff, pairs and glue and all kinds of things come up. And as ξ goes to zero, you have to find essentially everything. But there's a strong simplicity for the values of ξ between $\frac{1}{4}$ and 1, and this very striking simplicity is an important feature of the experimental results.

Finally, I mention in connection with this series of experiments the decay $\pi^0 \rightarrow 2\gamma$ when the experimentalists have been clever enough to measure not only the rate but also the amplitude. Without color, it would be in disagreement with theory by a factor of three in amplitude when we make the approximation that m_π^2 is small compared to $(1\text{ GeV})^2$. (It is, it's about 1/50 th;) and in the limit $m_\pi^2 \rightarrow 0$, you can calculate this decay through a theorem of Adler, Schwinger, Bell, Jackiw and others. This decay amplitude works beautifully with color; without color the amplitude would be wrong by a factor of 3 and the intensity of the decay would be wrong by a factor of 9. So this is a very good reason for believing in color, even though previously the other thing it agreed with was something it had been invented to explain, and was not therefore very persuasive as evidence.

Now we come to the stage where I should make a public apology to a defunct organization represented here called CEA (Cambridge Electron Accelerator). CEA, with its dying gasp, did an important and correct experiment, — subsequently confirmed by a team at SLAC, also represented here by Burton Richter, chief of the team, at higher energies. The experiment will be continued at still higher energies at super-SPEAR and perhaps at DORIS in Germany. This is to measure $\frac{\sigma(e^- + e^+ \to \text{hadrons})}{\sigma(e^- + e^+ \to u^- + u^+)}$. This should, theoretically, give the sum of squares if the quark changes: $3[(+\frac{2}{3})^2 + (-\frac{1}{3})^2 + (-\frac{1}{3})^2] = 2$. With charm 2 would become $3\frac{1}{3}$. This is at very high energies. Well, they've gone up to about 5 GeV and it is not anywhere near 2. It's more like 4 or 5 and still rising. It's something we don't understand. There are, of course, several possible ways the theory can weasel out of this. One is to say that if they keep going higher, it will go down. Well, they will go higher to about 9 GeV quite soon and we'll see if that's true, that it will start to level off and come down. They, of course, hope it doesn't. Another possibility is that some of the theoretical considerations involved in the unification of all the interactions imply that the experiments begin to involve other phenomena at energies like 5–9 GeV. If that's true, then those may explain the cross-section. The third hypothesis, of course, is the one that the experimentalists prefer, namely that the theories are dead wrong.

I don't have time to discuss heresies. I'll just mention briefly that there are some theoreticians who consider a variant of the theory in which above some very high threshold you can make colored things. This is quite possible. I shouldn't say that there is anything wrong with this. It's quite possible that this is a valid alternative. And even quarks can be made above some threshold — quark might actually be real particles. In that case, they probably have integral rather than fractional charges, but averaged over color to give the same fractional charges I quoted to you. There's the so-called Han–Nambu scheme in which the quarks have charges $1, 1, 0$ (for red, white and blue); $0, 0, -1$; $0, 0, -1$. The charmed one, if it's there, would be $1, 1, 0$. The charge averages are $\frac{2}{3}, -\frac{1}{3}, -\frac{1}{3}$ which is what we gave before.

That's a possible alternative scheme but I don't have time to say very much about it. In these alternative schemes, if the quarks are real, certain problems arise. Are they stable? If baryon number is conserved and the quarks have baryon number $\frac{1}{3}$ then the quarks would be permanently stable. We'd find them lying around, although perhaps in small numbers. They've been looked for in sea water, in oyster shells, and such things. People haven't seen them. But they might be lying around in small numbers. If so, they'd be very useful, I'm sure, for something. A new stable particle with totally new chemical properties will found some gigantic industry someday. You can imagine Route 128 lined with quark-onyx factories. Another possibility is that the quarks are integrally charged and unstable. But if that's true, then the proton is unstable. Here's another science fiction idea. The proton lasts 10^{13} years or so according to the experimentalists — 10^{26} maybe if you try a different decay scheme. Some people are playing with the idea that after 10^{13} or 10^{26} years elapse, the proton does decay. Well, maybe . That's a possibility. Another one is a picture invented by Nambu in which the quark–baryon number varies with color. He doesn't necessarily believe it but he was clever enough to invent it. In that case, the quarks could be ordinary baryons concealed in the spectrum of baryons already seen. They would be anomalous multiplets concealed in the regular baryon spectrum. That's a really weird idea. You have these choices if you abandon the normal picture of confined quarks. You may choose.

I want to end with a description of the kind of theory that people are trying to use to unify all these interactions. The idea is first to unify the electromagnetic, weak, and associated interactions; and then, maybe to unify those with a strong interaction and a lot of new ones which would come in only at very, very high energies. It's a rather ambitious idea. But the time looks propitious because these theories seem to have a lot in common. The kind of theory you have is one in which you have a lot of charges, which when commuted with one another finally form a closed system — they all commute to give members of the same set. You get a Lie algebra, and for each of the generators of this Lie algebra you have an intermediate boson: γ, $X^{\pm}$, Z^0, — we've discussed some of those. In the Lagrangian you start out with perfect symmetry: massless boson, everything perfectly symmetrical, perfectly gauge symmetrical. And then, *the vacuum*, the hero of the previous episode, enters here! It is unsymmetrical. There is spontaneous violation of almost all of these beautiful symmetries, creating these "minor differences" between the photon and the X^+. For example, the photon is massless and the X^+ coupling is left-handed. Things like that. All these things come as a result of the vacuum choosing some unsymmetrical configuration. It's actually not so implausible as it might sound, and you can construct mathematical models in which this really looks quite reasonable — not only mathematical models, but *serious* candidates for *serious* field theories that might actually describe all these interactions. Really. It's that prospect, that we might be close to writing down actual equations that really describe these things, that's so exciting, because it's that kind of general synthesis with real equations that we have been hoping for a long time. So you start with a theory in which all the leptons are massless and symmetrical, all the quarks have zero mechanical mass and they're all symmetrical, and all the spin one objects are massless and gauge invariant. Then, the vacuum-breaking comes in, smashes a lot of this, leaves electric charge conserved, probably baryon number conserved, leaves, for unknown reasons, leaves the photon massless and the neutrinos massless. But it would permit calculation of things like $\frac{m_e}{m_u}$ and $\frac{m_u}{m_s}$. Bjorken has already given us a candidate formula,

$$\frac{m_e}{m_u} = \frac{3\alpha}{\pi} \log 2 + 0(\alpha^2).$$

He hasn't exactly got a theory that gives this, unfortunately. But theories of the type we are describing may very well lead to a formula like this, and in fact might even lead to this one. It works to a third of a percent and it's a very nice candidate formula, because if the muon is the one that gets the mass originally from the vacuum breaking, and the electron gets its mass from an electromagnetic and weak correction, or a correction from a unified theory including electricity and weak interactions, then you'd expect the factor α. Radiative corrections always come in with π in the denominator. Viki discovered this as a young man when, as he once told me, he estimated the scattering of light by light amplitude and got all the dimensional factors and the powers of α correct. When it was subsequently calculated exactly by Euler and Kopple, the result was the same except for a numerical factor of $360\pi^4$. If one is careful with $\hbar$'s and 4π's in the α, and if one always remembers to put in π in the radiative correction, then, such things perhaps can be avoided. Not by me certainly — I probably would have gotten the $360\pi^4$ in the numerator instead of the denominator. The factor 3 in Bjorken's formula looks pretty nice also, and there ought to be a logarithm of something. The only thing you have to do is dial the argument of the logarithm, and you find that when it achieves the value 2 you get excellent agreement with the observation, and just the right error of a third of a percent to be compatible with $0(\alpha^2)$. So I myself think this will be the right formula, but it hasn't really been derived yet.

The strong interactions may be described by a very similar theory. In fact, here there might not be any breaking. It's conceivable that the strong interaction theory is just a plain, Yang–Mills, gauge theory of quarks and color octet gluons. It's conjectured, but it hasn't been shown that in such a theory the $\frac{1}{r}$ potential that arises from these fictitious but have massless gluons might be converted into a potential which varies $\sim r$ or $\sim r^2$. The "bag" men at M.I.T. work with something very much like this but they add a bag to make absolutely sure the colored things stay inside, not having faith that this $\frac{1}{r}$ will turn into an r or r^2. Time will decide which is right whether it's necessary to introduce the extra mechanism or not.

Now let me ask some questions at the end, open questions. Is there really charm or is there some new explanation of the absence of the neutral $|\Delta S| = 1$ currents? The smallness of the $K_1^0 - K_2^0$ mass difference? This can be investigated experimentally, especially by looking for charm. It should show up very readily if looked for in the next year or two. Is the algebra of the electromagnetic and weak and associated interactions really just the simplest, smallest one with γ, $X^{\pm}$, Z^0 or is it something bigger? If you make it bigger, will you in fact unify the electron and the muon, the various quarks and so on, and get something in agreement with observation? For the strong interaction, is the Yang–Mills theory the plain ordinary gauge theory of color octet gluons binding quarks? Is that good enough, or is it necessary to modify the gauge theory with a bag or some other kind of ball or stuff that's added into the theory separately? We don't know which. (I would call that a modified gauge theory.) Are the quarks standard and confined with their fractional charges never coming out, or is one of these heretical pictures right, with integral charges and with quarks emerging, and stable quarks or with decaying protons? What about the vacuum breaking? Do we really understand how the vacuum symmetry gets broken? To do it in one of today's field theories is the only ugly feature of the theories being proposed today, because it requires a vast number of these scalar and pseudo-scalar fields which assume non-zero expectation values in the vacuum. Also, several new parameters have to be introduced whereas previously there was only one parameter for a given group, only one coupling constant. But as soon as you introduce these vacuum breaking explicit fields, then you get extra parameters and a lot of different fields. So some people hope that maybe these scalar and pseudo-scalar things that assume non-zero values in the vacuum can be derived dynamically from the rest of the theory and will not have to be put in explicitly. That might be a big help if someone can do it, and show that the theory is still finite. This, by the way, is probably not the same scalar field that T. D. Lee was talking about. He was talking, if I understood correctly, about an effective field ϕ for something like a quarkless meson. This would not be one of these fundamental vacuum breaking fields entering into the basic equations of the quarks and the leptons.

Anyhow, the success of what has happened so far, and the prospects opening up, suggest that there is some hope for a general theory in which the world is made of quarks and leptons, coupled together by gluons, photons and so forth, all in some sort of gauge theory, and with this whole mess coupled to gravity. So we have spin one-half particles, quarks and leptons, as the basic constituents of the entire world, bound together by spin one bosons, gluons, photons, $X^{\pm}$ and so on, some of which can get out, some of which can't, and Einstein's gravity coupling to everything. Nobody knows whether this will work but there is the smell of a synthesis in the air.

7

What Angular Momentum Can Do to the Nucleus*

Ben R. Mottelson
Niels Bohr Institute, Copenhagen

Viki, I bring you the warmest greetings from all of the group in Copenhagen. I also would like to take this occasion to tell you how much I personally treasure the discussions that we've had through the years. They stand for me as inspiration and also as something that has made physics more fun, more meaningful; and I thank you. I also want to tell you about some of the new developments in nuclear physics, hear your suggestions and criticisms of some ideas that we've been discussing.

As you may know, the facilities that are just coming into use these days produce accelerated beams of heavy ions, which in collisions with other heavy ions create systems with several hundred units of angular momentum. And thus we envision a major expansion of the whole field of nuclear structure studies associated with the possibility of studying systems in which the centrifugal forces are comparable to the total cohesive forces that hold the system together. In this whole new field, one of the first questions you may ask is, "What is the largest angular momentum that the system can accommodate?" You may, of course, first immediately think of the centrifugal forces acting on each single nucleon in the system and for sufficiently large angular momentum these forces must become large enough so that the single nucleons become unbound and are ejected from the system. But such a limit you can quickly convince yourself is too permissive. Instabilities

---

*Editor's note: Dr. Mottelson has not read this manuscript, which was transcribed from a tape of his talk. The content represents work done jointly by A. Bohr and B. Mottelson.

come before that; instabilities associated with collective deformations which lead eventually to fission. The main terms in these fission instabilities are expected to be associated with the balance between the macroscopic deformation energies and the rotational energies of the system and thus we are led to look at the energies of a rotating, liquid drop. And so that's the first system that I would like to say a little bit about. The energy will then be comprised of the surface energy which will be a function of the deformation, the Coulomb energy, and a rotational effect that expresses the dependence of the energy on the angular momentum. The rotational energy, we may expect, will correspond to that for the rotation of a rigid classical system as soon as we come to angular momenta that are so high that the nuclear superfluidity is destroyed. We'll come back to some aspects of that problem later. This leads us to expect that we have a classical rotational energy term with a moment of inertia I, corresponding to rigid rotation. In order to get some order of magnitude feeling for the sizes of the different terms, we may compare the rotational energy at zero deformation, the rotational energy for a sphere, with the surface energy, also for a sphere. The rotational energy then involves the square of the angular momentum and the moment of inertia involves the total mass of the system and the radius squared. So there is a five-thirds power of the mass number A there. Then in the surface energy, which depends on the area and the surface tension constant, there is a two-thirds power of the mass number. So the characteristic parameter then is $(I^2/A^{7/3})$ and the magnitude of the angular momentum we may anticipate will be of the order of $A^{7/6}$. For an angular momentum of that order, the rotational energy is comparable to the surface energy and we may expect the fission instability to set in. This is, of course, neglecting the effect of the Coulomb energy. As we go to heavier nuclei where the Coulomb energy becomes important, we will get a stricter limitation.

We now ask what happens to this liquid drop as we begin setting it slowly into rotation. Well, for small values of the angular momentum we may expect that we will get equilibrium deformations that are of the order of this $(I^2/A^{7/3})$. The fact that the moment of inertia has a linear term in the deformation means that the system begins to deviate from spherical symmetry achieving an oblate shape just as the rotation of the earth distorts the system into an oblate shape. The magnitude of such a deformation is very much less than that associated with shell structure which we'll come back to later. But, it increases with angular momentum and eventually must be the main deformation effect causing an increasing oblate-shape deformation with the angular momentum.

However, soon an instability in this oblate shape sets in. This instability is known from the study of rotations of astronomical bodies and corresponds to the change from the MacLaurin spheroids to the Jacobi ellipsoids. The idea here can be very simply seen if we look down from the top on this deformed rotating system and we ask what is the effect of a small deformation away from axial symmetry of the system. The effect of such a deformation is measured by an amplitude, γ. The moment of inertia then receives a term of order γ^2 since we move an amount of mass, γ, through a distance, γ. This is the same order of magnitude as the increase in the surface energy and therefore, if the angular momentum is small, the surface energy will be larger than the rotational energy, and the system will be stable with respect to such a deformation. But with increasing angular momentum the rotational term which is proportional to the square of the angular momentum times γ^2 must eventually become bigger and therefore our system will become unstable with respect to this deformation. We make then the transition from the axially symmetric shape to the triaxial shape. There is, of course, eventually a tendency to form a string rather than a planar system. So we may call this the sheet-to-string instability.

If we are comparing this sequence of equilibrium shapes with the saddle-point shape, then the important point is that the saddle-point shape is a very extended configuration compared with the shapes that we've been considering along the equilibrium line. Being extended, the moment of inertia is larger, the rotational energy is smaller and, therefore, eventually the rotational energy of the saddle-point shape will approach that of the equilibrium sequences and the system will become unstable with respect to fission. The saddle-point shape is the shape of the nonrotating system. It's the shape in which the surface energy is in balance with the Coulomb energy which is a destructive energy trying to drive the system to larger deformations and therefore the energy of the saddle point is the activation energy for the fission process. So we get a picture that looks something like Fig. 1 in which this sequence of shapes that I've just been discussing runs along and becomes unstable with respect to the triaxial shapes. The saddle-point shape has an energy which for low angular momentum may be quite large but it increases slowly and eventually intersects with the equilibrium shape and then we have the fission instability.

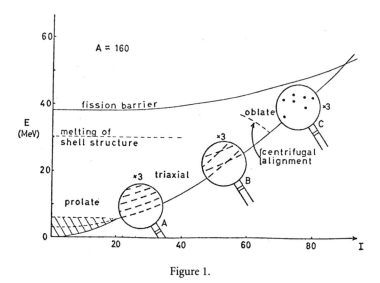

Figure 1.

Well, this is the picture for a single nucleus, for a particular value of A. We may look more generally at this instability for the whole periodic table and plot the maximum angular momentum that we find, the point of intersection between the saddle-point shape and the stable sequence, as a function of the mass number of the system and we get a curve that looks like Fig. 2. The dotted line represents the point for instability with respect to the triaxial shape, so in this region we have triaxial shapes while in the lower region we have oblate shapes. Outside these regions we have the unstable system. The outside curve represents the point where the fission barrier is zero. I've also sketched a curve which gives the point at which the excitation energy for fission, the difference between the ground state and the saddle-point shape energy is equal to 8 MeV. That is the approximate dividing line for those systems which will decay by fission, and those that will decay by neutron emission. For systems with lower angular momentum than this, the most important decay mode is that associated with neutron emission. While for those above this point, the most important decay mode would be fission.

Before I compare the limits sketched in this picture with experimental data we should, of course, examine the most important quantal corrections, or quantal limitations, to this classical macroscopic analysis that I've sketched here. And there, of course, we must pay special attention to the

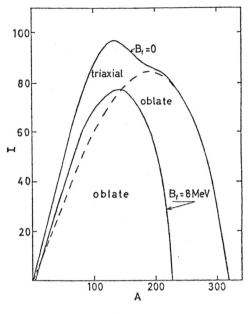

Figure 2.

single particle effect of the shell structure of the nucleus. The order of magnitude of shell structure effects can be rather simply understood. The shell structure reflects the quantization of the independent particle orbits in the whole nucleus. The eigenstates of that motion, the single particle motion, represent an irregular sequence and these irregularities can be roughly approximated by the shells that then have a characteristic separation that I'll call $\hbar\omega$. So the total energy that's associated with this irregularity in the spectrum of the single particles is the shell structure energy which is of order, (the number of orbits in a shell) x (the energy separation of the shells). The number of orbits in a shell is $A^{2/3}$. The separation between the shells is $A^{-1/3}$ so this energy is fundamentally of order $A^{1/3}$. In a typical heavy nucleus like the one that I showed, this would be something between 5 and 10 MeV. It's, therefore, a small energy compared to the main terms I wrote down before. I remind you that the maximum rotational energy is of the order of the surface energy which is of order $A^{2/3}$ which, in the example we just looked at, is something more like 50 MeV. So this shell structure energy will be something that may be important for the details but is unable to change the basic order of magnitude of the estimates that

I made before. Similarly, a very significant quantal effect in the low lying states is associated with the condensation into superfluid that characterizes nuclei in their low lying states. But there also, we can estimate the energy associated with the pairing as in the estimate for a superfluid. The condensation energy is the square of the binding energy of a pair divided by the separation between the single particle levels at the Fermi surface. The binding energy of the pairs, goes approximately as $A^{-1/2}$. The separation between the single particle energies is, of course, A^{-1}, so we get something here which is A^0 which is a few MeV in almost all nuclei and, therefore, also, something that may be important for the details but cannot change the basic trend in the data we looked at before. So we may think it makes some sense to compare the curve that I showed a few minutes ago of the maximum angular momentum with the estimates based on the rotating, liquid drop.

The next figure, Fig. 3 shows the way the experimental data that is already available provides some interesting tests for these limitations. The figure shows cross-sections for a collision involving argon on silver at about 30 MeV. Plotted are the cross-sections as a function of the mass number of the products that are produced. The group at about mass number 130 can be identified with nuclei in which the colliding ions have formed a compound nucleus and then evaporated of the order of 13 nucleons surviving fission

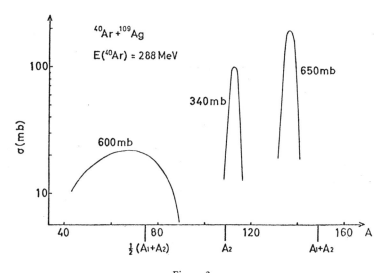

Figure 3.

at each evaporation stage and then finally came out as nuclei with mass numbers about 13 units less than that of the sum of the target and the projectile. The next group of nuclei, a broad group around a mass number about half that for the compound nucleus, can be identified with those systems that form a compound nucleus but fission somewhere along the evaporation chain before coming to form a stable nucleus. The third group that's shown, with a mass number just a few units greater than that of the target, correspond to grazing collisions in which the projectile just went by weakly interacting with the target and added to it, apparently, just one or two nucleons. There must, of course, be another group corresponding to the projectiles that went by and lost the one or two nucleons and would come at lower mass numbers that have not been taken into account in this figure.

The cross-sections that are measured for this collision process provide an immediate test of the angular momentum limitations that I mentioned before. A test that is independent of any detailed discussion of the mechanisms of the collisions with the target. If we consider collisions with impact parameter, b, then wc get cross-sections πb^2 and the angular momentum that are involved are simply the incident momentum times the impact parameter divided by $\hbar$. So simply from the cross-sections we can deduce the minimum impact parameters that must be involved and from those the angular momenta. The magnitudes of the cross-sections in the figure tell us that the group that went into the compound nucleus and survived fission involve angular momenta up to about 80 units. And that, in fact, agrees quite well with the prediction. If we add together the compound residue group and the fission group, then we get a total cross-section corresponding to angular momentum of about 108 units and that agrees to within about 10 units of the estimate. I don't want to emphasize the quantitative agreement here because these experiments are new. There are experimental uncertainties, and that's not the point. The main thing I would like to emphasize is that the tools are in hand for making a detailed study of the limits of stability associated with angular momentum and that the preliminary evidence we have indicates that the order of magnitude of the numbers that appear here is not too bad. We are able to produce and study systems of large angular momentum.

So that completes the first chapters of our discussion, gives the boundaries of the nuclear world as defined by angular momentum. What kind of effects are we likely to find within that domain? What is the effect of the centrifugal forces on the nature of the systems that are produced, on the structure of the nuclei ranging from angular momentum zero up to these maximum limits? For that question, we can make especially deep studies on the basis of the quantal states in the neighborhood of the lowest state of given angular momentum. This lowest state of given angular momentum is called by the nuclear physicists the Yrast State, and by the particle physicists the leading trajectory. "Yrast" comes from the Swedish word for dizzy (similar in Danish and old Germanic). It's the superlative form, you see, "ast" corresponds to "est" in English, so it's the most dizzy, confused or spinning state in the nucleus. I don't know whether Yrast is a good word, but we need a word to talk about this state and the states in this region and that's the word I'm going to use in the rest of this lecture.

So in the neighborhood of the yrast line, in the yrast region, the nucleus can be considered cold despite the fact that it may have a very large excitation energy of the order of 50 MeV. But all this excitation energy is tied up in generating the angular momentum and, therefore, the system is cold. We can expect the density of states to be similar to that in the neighborhood of the ground state and the motions in the system to have a similar high degree of order. And so detailed spectroscopic studies give us the chance of probing in considerable detail the structure of the system. We can expect, as we go along the yrast line from the lowest angular momentum to the greatest, to see a sequence of different phases of nuclear matter that reflect the changing interplay between the macroscopic centrifugical deformation effects that I've been discussing so far in terms of the rotating liquid drop, and the quantal effects associated with the motions of the independent particle in the system. I'd like to briefly go through some of the properties of those phases and the things that we might anticipate as we carry out these studies of the spectra along the yrast line.

For the lowest angular momentum, if we start out in the region of angular momentum up to around 20 units, there is a great wealth of experimental data and we know quite a bit about the structure of the systems. I'll briefly remind you that there the shell structure is the dominant term. The macroscopic rotational energy that I mentioned before is a relatively minor effect. The shell structure is important especially in its effect on the shape of the system. If the system corresponds to closed shells, then the spherical shape has a special stability as in the spherical shapes of the inert gas atoms associated with their closed shells in atomic physics. However, if we have a number of nucleons outside of closed shells, there is a degeneracy in the states available to them. This degeneracy then opens the possibility for spontaneous symmetry breaking. A large class of nuclei exploit that possibility and exhibit shapes in which the deformations deviate from spherical symmetry by an amount which is given by the number of particles outside of closed shells divided by the total number of particles. The number outside of closed shells can be at most $A^{2/3}$ and the total number is A so we get $A^{1/3}$ as a characteristic magnitude for the deformations associated with this spontaneous symmetry breaking. As always in spontaneous symmetry breaking, there is a collective mode and the corresponding collective mode here is that of rotation. For almost all of the nuclei that are known this spontaneous symmetry breaking leads to shapes that approximately correspond to prolate ellipsoids. The rotational motion is about an axis perpendicular to the symmetry axis. Such a rotation, of course, is then associated with a very strong time dependent electric field outside the nucleus and, therefore, these states have very strong electric quadruple transition probabilities linking one state to the next. So we get a regular sequence of collective rotational states linked by very strong electric quadruple transition probabilities.

The pairing is also very important for the structure of the rotational motion of these nuclei. The pairing was, of course, identified in a very early stage of nuclear physics in terms of the characteristic difference between the odd and even masses. But it was later recognized as a systematic correlation effect very similar to that which is responsible for the superconductivity of electrons in metal. We can think of nuclei in their low lying states as small drops of superfluid. Now if we were to consider the rotation of a macroscopic drop of superfluid then the collective motions must be those of a potential flow and potential flow can only take place in a simply connected domain if the surface moves. Therefore, if we had a spherical superfluid drop, the only way to carry angular momentum, to exhibit a collective flow, is to destroy the superfluidity along vortex lines. And as we know from the case of liquid helium, the vortex lines distribute themselves throughout the volume of the system in such a way that the average velocity field is like that of rigid rotation. However if such a superfluid is deformed, there's also a possibility to have a collective rotation in which the surface moves and corresponds to a kind of surface wave with an amplitude proportional to the deformation of the system. The moment of inertia, then, goes as the square of the deformation of the system and is, therefore, very much less than that for the rigid rotation. Now, for the nucleus, the radius of a vortex line is determined as in the superfluid by the coherence length and the relatively small magnitude of the binding energy of the pairs implies that the coherence length is very large and larger than the nucleus. And, therefore, the collective motion associated with nuclear rotations is intermediate between that of potential flow and the rigid field that would be characteristic of a very large system. We then have a characteristic of the superfluid phase that we encounter in these low energy systems — the moment of inertia is less than it would be for a rigid rotation, typically by a factor of two. This, of course, immediately means that as we go up along the yrast lines and increase the angular momentum of the system, the energy of the superfluid phase is rising more rapidly than the energy of the normal phase where the rotational moment of inertia would be that of rigid rotation, and, therefore, the rotational energy is less. Eventually there must be an angular momentum at which it becomes more favorable to give up the condensation energy of the pairs and go over to the normal phase and give up the superfluidity. If you use the parameters that are known, you come to the estimate that this occurs at 15–20 units of angular momentum. So we expect that our picture is characterized by a superfluid region and the normal phase outside that region. In fact, the experimental studies of the spectrum of states in the yrast region gives rather strong indication that something like that is happening in the region where it's expected.

The way we analyze the energies along the yrast line to exhibit the occurrence of this transition is to plot the energy as a function of angular momentum along the yrast line. But we like to exhibit this result in terms of the moment of inertia as a function of the rotational frequency and we do that by essentially interpolating between the discrete values given by experiment so we can differentiate and then we can obtain the rotational frequency as simply the derivative of the energy with respect to I, and the moment of inertia as the ratio between the angular momentum and the rotational frequency. On that basis, the observed spectra indeed exhibit a remarkable singularity in the neighborhood of the expected region. When we plot the moment of inertia (defined as I just indicated) as a function of the rotational frequency, it first increases smoothly as a kind of centrifugal stretching effect until we come to angular momenta of the order of 12 or 14 where a sharp variation sets in. I should confess that we don't have a quantitative description of such a phase transition as expressed in terms of the discrete quantum states of the system and so this data is not quantitatively interpreted. But I think the indication is very strong that we should identify this singularity with the approach to the normal phase especially since after the transition we come to a moment of inertia which is approaching rather closely that for rigid rotation. I should say that it is probable this is not the only thing that is happening here. There is also an indication that in the same region the effect of the Coriolis force on individual orbits is also playing a role. But until we have a more systematic description of the way such a phase transition occurs in terms of the discrete quantum states of the spectrum, it is difficult to give any detailed description of this data or of the immediate region above it in angular momentum.

As we go further along the yrast line, there is a constant competition between the macroscopic distortion associated with the centrifugal effect I spoke of earlier and the shell structure effects associated with the single particle motion. But because the macroscopic distortion effects are continually increasing with the angular momentum, we must expect that sooner or later they will dominate the shell structure effects. The deformations associated with the centrifugal distortion become large compared with the $A^{-1/3}$ that characterizes the deformations in the lowest states in the system. And so we expect that eventually the centrifugal distortion drives the system over into the oblate shapes that are the characteristic effect of the rotating, liquid drop configuration.

But if the system acquires such an oblate shape, we are then in a regime that's totally different than the one I've just been discussing. In such an oblate potential the motion of each individual particle has, as a constant of the motion, its angular momentum along the symmetry axis. The angular momentum along the symmetry axis is constant for each particle moving in such a potential. Therefore, if we want to generate a large angular momentum along that axis, an angular momentum that may be 50 or 60 units, the only way to do it is to align the momenta of each of the individual particles. The lowest state would be obtained then by analyzing the one particle spectrum in that potential and finding the easiest way of occupying those states so that the resulting sum of the contributions from each quantum state add up to the total angular momentum that the system needs at that point. Then, if we want to go from one state to the next along the yrast line, there's no sense of collective rotation. We find the next state by finding some other way to occupy these single particle orbits, taking particles from some of the ones that had previously been occupied and occupying other ones with larger components of angular momentum along the axis. So the whole spectrum along the yrast line has no family relationship. There are very large fluctuations in the nature of the states so there's no collective transition between the separate states. The transition matrix elements are those associated with transitions of a single particle and, therefore, are often subject to selection rules. They may even involve the transitions of two particles and, therefore, a vanishing transition matrix element in the first approximation. In any case we never get a matrix element larger than that of a single particle transition. You can see the same thing from the classical picture of the system. For an oblate system rotating with its angular momentum along the symmetry axis, there's no time dependent electromagnetic field and, therefore, no strong radiation.

So for that system we expect great fluctuation, no large matrix elements, and quite often the possibility of strong selection rules, making the transitions quite slow. Therefore, we may imagine that in such an oblate regime the occurrence of isomerism may be quite frequent. Some of the yrast states there may have no chance to decay to a lower lying state with any collective or even with a single particle strength. Well, I can only say that the discovery of such isomers would be a remarkable phenomenon that would open enormous possibilities for the detailed study of the yrast region. You can imagine a state at 30 MeV of excitation energy living for days or years or maybe only microseconds but still so long that it can be clearly separated from all the other systems. It would then decay by a long cascade ($\sim$30) of gamma rays. That's really a remarkable phenomenon and it would tremendously enrich our knowledge of the whole spectrum of states and open the possibility of probing deeply into the nature of the different structures.

These conjectured states along the yrast line have sometimes been referred to as superdizzy nuclei. Certainly if they're found, they'll be a very interesting phenomenon. The interesting expectation is that at the point where the instability to triaxial shapes sets in, the system cures itself of this superdizzy disease. The systems with triaxial shapes at higher angular momentum will then have regular rotational band structure because they are executing rotation around an axis where there is no axial symmetry. The region from the destruction of superfluidity up to the region where the oblate shapes occur may also be characterized by shapes with a triaxial symmetry. It will be a very interesting region because the triaxial symmetry means that we have rotational motion of a body with three different axes and would get the characteristic rich spectrum in each rotational band. The system must change over from one shape to the next as we go along the yrast line in order gradually to adjust itself to the distortion that's being imposed by the centrifugal forces. But we don't have pairing, we don't have superfluidity which insured that there was a unique ground state. The only way these changes in shape can occur are by changes in configuration. So we have systems of rotational bands intersecting and crossing through each other along the yrast line as the shape gradually adjusts to the centrifugal distortion, then changes over to the oblate system with its remarkable irregularities and then finally perhaps again to another system of triaxial shapes.

I don't have to say that this picture that I've sketched contains an enormous speculative element; there's a lot here that's very uncertain. But I do want to leave the impression that we see here a very rich and very exciting field. I am sure, we're going to learn an awful lot about the effects of angular momentum on nuclear structure and on the study of the way in which transitions between these different phases of nuclear matter are expressed in terms of the discrete quantum states of the system.

8

Physics for Mathematicians

S. M. Ulam

University of Colorado, Boulder, Colorado

When I was asked to give a talk here, being just a mathematician among the distinguished array of physicists invited to speak, I had great hesitation. Then it occurred to me, if Viki Weisskopf can conduct a symphony orchestra, maybe I can talk about physics. I felt consoled until yesterday evening when I discovered that he is a professional, and so I feel very, hesitant again. My title, "Physics for Mathematicians", will almost mean physics without mathematics. My interest is really to paraphrase a famous statement, not what mathematics can do for physics, but what physics can do for mathematics. That is the underlying motive.

In the last decades, there has been a widening gap between theoretical physicists or people who know physics and the professional mathematicians. Perhaps that is unavoidable because within these sciences themselves, there is increased specialization. However, it occurred to me that there were mathematicians in the 19th century who were at the same time physicists or who contributed very significantly to physics. Think, for example, at the end of the century or the beginning of this century, of Poincaré whose marvelous popular books on science in general and physics and mathematics especially, had such tremendous influence on the young people who read them — say in the 20's and could see the inspiration which both subjects derived from each other. And I noticed that Viki's book, or rather his article on the impact of quantum theory on modern physics starts with a quotation from Poincaré. The quoted lines were written about the time that Viki was born, which was not so terribly long ago; and if you read a book by Poincaré which I recommend even now for historical perspective, you will see how much in this period of time has happened.

Now, what I want to talk about is not the, so to say, service which the mathematician can render to physics, but about the conceptual possibilities which even now, or perhaps more than ever now, the historical spectacle of physics provides to mathematicians, even the most abstract ones. And at the same time, I will mention just very vaguely a few mathematical ideas which may perhaps play an important role in the physics of the future, near future, or perhaps more remote future. But, of course, that is always hard. As Niels Bohr said, it very hard to predict, especially the future. Poincaré was one of the last universal minds of this sort. In more recent times, one could mention Herman Weyl and to some extent, John Von Neumann. They contributed more than mere grammar to the science of physics; their points of view influenced the physical thinking itself. What I will do is mention several areas of what we call physics which might stimulate mathematicians and here and there mention some mathematical thoughts which have influenced at least the techniques of physics.

One such field known to perhaps everybody is group theory. At first, one might have thought group theory merely played the role of a very good filing system, but it has assumed an important fundamental role. Here I come back to Poincaré's books. There are four books; they are very popular and are on the very highest level, both from the point of view of literary merit and philosophical merit. They are called *Science and Hypotheses, Methods of Science* (the one I want to refer to), *The Value of Science* and *Last Thoughts*. I think the one I'm mentioning now was written around 1908 when Viki and some others of us were born. It contains some of his speeches at the World's Fair in St. Louis in 1904. Among them is a very famous address on the unsolved problems of mathematics and physic where he discusses the discovery of certain harmonies of nature as one of the aims or one of the most important stimulations for work in physics. In this word "harmony", he points out, are more general ideas than the idea of a group. It embodies all the analogies, formal, obvious or hidden in various parts of physics. To discuss harmony Poincaré used the, by now so familiar, role of linear partial differential equations in several apparently quite separate and different parts of physics. It is really very strange to look at this book now and compare the picture which one had of the world of physics, of nature, with what is now known. And there are some curious little items. I noticed he mentions the

Japanese physicist, Nambuma, who around 1900 or a little before, certainly before Niels Bohr, proposed that the atom is one big positive electron surrounded by a ring of negative electrons and draws the conclusions which, of course you all know, later led to Bohr's model.

The ideas of Poincaré had a different impact on the development of mathematics from those which Hilbert formulated in his famous lecture to the Paris Congress around 1900. Indeed Poincaré himself at the same congress gave a lecture on the different mental attitudes within mathematics itself. He said there are two types of mathematicians. There are the analytic ones who given some axioms or rules like to produce new ones with utmost ingenuity, or like to dissect and construct new objects of mathematical thought. And there is the other type, the intuitionists, who, without being so much interested in the formal aspect, divine by analogies, the pattern of the world of thought. Clearly such distinction must exist among physicist. There are people who use as virtuosos, the mathematical techniques; and there are others who stress more the concepts in physics, and I would think Viki belongs, if I can judge at all, to this intuitive type. It is this intuitive type which is *prima facie* useful to mathematicians who are curious about the physical world.

The situation in mathematics as I said before has led to increased fragmentation and specialization. By now, there are ten or twelve different kinds of topological methods or fields, and often the practitioners of one do not know much about another. Once, last summer I was asked to give a talk at some Jubilee or memorial; actually, it was a 25th or perhaps 30th anniversary of the computer in Princeton. And as I was talking in general terms about problems of mathematics which can be studied by heuristic experimental work on computers, it occurred to me to sort of estimate the number of theorems which are proved yearly. What is a theorem? Well, it is something that says "theorem" and is published in a recognized mathematical journal. I quickly tried to multiply the number of journals by the number of issues by the number of papers in each issue and by the average number of theorems and when I got the answer I thought I had made a mistake. I said one hundred thousand theorems are proved yearly and my audience gasped and I was sort of appalled myself and thought I must have made a mistake of an order of magnitude. But the next day two younger mathematicians who were in the audience came to me and said they had made a better search in the library and they found out that the number was closer to two hundred thousand yearly. Will that sort of thing happen in physics? Will there be again proliferation and fragmentation? There is much less danger of it. There is a certain unity in physics and even physicists who specialize in this or that field are aware of others. Of course, in physics there is a proliferation of publications, too. Somebody told me that if one extrapolates the number of pages in the *Physical Review*, there would be so many that even at light velocity you wouldn't be able to peruse them all. Physics, however, still keeps its certain unity. Physicists, no matter how specialized they are, on the whole have interest in and knowledge of the foundations of physics which is not at all the case in mathematics.

To start telling you about some more definite things, I would say one problem that would interest a mathematician now, a mathematician who is familiar with the ideas of physics, would be the problem physicists as such don't consider a physical problem. The problem roughly would be this: Is the world fully describable by physical laws? Is it finite or infinite? I mean it in the following sense. Historically there was always, since the Greeks, the idea of atoms which are indivisible and unanalyzable. During the last, let's say fifty years, the molecules were found to be composed of atoms, atoms in turn of nuclei, and electrons. The nucleus itself was a group of nucleons and now, as you know, people take a little bit more seriously the picture of a single nucleon being composed of partons, perhaps quarks, perhaps something else. The mathematician's question is: Will this go on, perhaps forever? It is a question which, until recently, you could have said is merely philosophical. But perhaps less so now. If the things are very small and in a certain sense, throw a shadow ahead of them, perhaps it might be possible sometime in the future to consider the world infinite in structures going on through stages which are not at all necessarily similar to each other. The same question was well considered long ago in the large: Is the universe of stars, galaxies and the groups of galaxies bounded? Is the metric elliptical or is it hyperbolical or is the universe truly infinite? And this to some extent, can be investigated or guessed. Perhaps it is easier in both cases to assume the universe is infinite. These are the models which would delight the pure mathematician. Mathematicians like infinities. Physicists are beginning to wonder whether the subdivision of structure will continue forever. I will say this is, as yet, not a problem which one can precisely investigate or even devise experimental tests for. It is more convenient to work with the calculus of infinitesimals than with finite differences in pure mathematics. Is there an analogous situation visible in the next ten or twenty years in theoretical physics? If so, then, a very interesting and, to some people, ominous possibility will appear. As most of you know, in mathematics in the last thirty years, starting with the discoveries of the logician Gödel, it was found that any finite system of axioms or even countably infinite systems of axioms in mathematics — if not trivial — will allow one to formulate meaningful statements within the system which are un-decidable, that is to say, within the system one will not be able to prove or disprove the truth of this statement. Fortunately, I think one is very far from that yet in physics, but it is permitted to speculate whether the number of essentially new phenomena, the number of lows will increase forever, or as was the great hope of the 19th century and even of the first part of this century, there will be a few ultimate laws which will allow one to explain the external world.

Psychologically, there is a difference between mathematicians and physicists, but I think it might attenuate in the following sense: Mathematicians start with axioms whose validity they don't question. You might say it is just a game — "the great game" as Hilbert called it — which we play according to certain rules, starting with statements which we can not analyze further. Now to some extent, at least, I would claim that in physics the situation is inverse. Given a lot of facts, let's call them theorems, we look for axioms, that is to say physical laws, from which they would follow. So physics is an inverse process. And in mathematics itself, you could think of such a game: Given some theorems in a certain well-defined notation or algebra, find the underlying axioms. This game has not been played and I'll tell you why; it is because the idea of algorithm and the formalization of what we call definite equational theory is very new. So mathematicians have never hesitated to create objects of their own. Right now there begins to appear some work which I would consider of this inverse type, perhaps it was stimulated by the existence of computers. The questions for a mathematician is how to approach this vast spectacle of physics, whether it is really a game starting with given laws or whether on the contrary, as seems to be the case, given an enormous number of facts or classes of facts it is to discover the rules from which they follow. It is, at least some physicists think so, the question of formal beauty. I would rather not subscribe completely to what Dirac said several times — first write beautiful equations and then if they are really very nice discover that they are probably correct. He may just have stated this extreme point of view in order to encourage this kind of work.

He himself had one of his greatest successes that way. This should interest mathematicians who are logic or set theory oriented. Let me tell you, by the way, that one could say that even the foundations of mathematics itself are written on sand. The recent discoveries in set theory show that the intuitions which were so commonly shared among mathematicians about the degrees of infinity, are dependent on more assumptions than we had imagined and to an extent vary from person to person.

Physics, however, is a world in the singular. In mathematics there is not just one geometry, as there was in the beginning with Euclid, there are many geometries. Just as there are many geometries, who knows, there might be several different types of physics. Not merely different in formulation as the same thing in different languages, but given for instance different situations (call it other universes), the laws could be different. Not only the constants could change in time but conceivably just as a mathematician's plaything, it could be that some of the rules or laws themselves are variable. This could be stimulating work for mathematicians interested in the world of foundations, My own taste would not run to that at all because it is a *post mortem* activity. It is again a question of taste.

I come to my next topic. What is the stage on which physics is played? Well, it is the structure of space and time. To me, it is rather surprising that the idea of space — Euclidean or Newtonian space — was generalized by mathematicians in literally thousands of papers and books, to very general topologies, very general metrics, while very little was done in a similar vein with the four-dimensional non-positive definite metric of the Lorentz space. There are hardly any speculations on how to generalize space-time and play with it mathematically. I'm not saying it is or will be important in physics except as a mathematical exercise or curiosity. The classical ideas of Riemannian surfaces are only very, very special things. Nowadays, mathematicians consider strange topologies in the small which are non-Euclidean even locally, and one day again, it wouldn't do any harm if somebody could write an article on this similar to the popular lectures of Poincaré even if it is much more difficult. It could stimulate or indicate the possibility of using, at least formally, the idea of a space full of holes within holes. Such things were considered long ago by the creators of set theory and topology. In other words I would say that geometry itself is the stage on which physics is played. I have myself written several "propaganda" articles on these possibilities.

To tell you the truth, I was just out of high school when I read about the new quantum theory. It surprised me that the ψ-function was defined for all points *xyz* and that *xyz* were defined with infinite precision. Of course, the difficulties with alternative possibilities are well-known. There is no minimal distance possible which would be consistent with the Lorentz special relativity transformation. But it need not be that there is a minimal distance. Just like there are sets between points of which the distances are arbitrarily small but not all of which "exist", only some of the points "really" exist, the other being "virtual".

I have myself written, together with some other mathematicians, little notes on "p-adic" time-space in which other groups would play the role of the Lorentz group. What would be the equivalent of the light cone, etc.? These are sets of points which have the "time" norm equal to the "space" norm. Indeed listening to Professor Mottelson's lecture, I was struck by how much the classical ideas help in the realm of phenomena where the quantum theory was supposed to be complete and at variance with any sort of mechanical model interpretation. This is the inverse correspondence principle — the classical ideas seem to have validity, at least as models of stimulation, for the very small. Of course, it is very hard to make this transition to the macroscopic phenomena, to sew it together without the seams showing. That is the problem. I remember Fermi's frequent visits to Los Alamos. On walks I would tell him the impressions of a young mathematician about physics and how the derivations weren't satisfying for some people. I told him that when I first learned, just out of high school, about the success of the Schrödinger equation, and how pulled out from thin air, it gives the spectral line values with accuracy to three or more decimals. I would have considered it a fantastic success if it was correct to 10 percent. Fermi said, "You know, Stan, it has no business being that good." By the way, sometimes in discussions about some problem he would tell me, "Oh, now they say that one should make the following assumption." I was very much amused because, "they," of course, included Fermi, himself, one of the creators of the quantum theory. Here is a problem related to the nature of what we call space-time, and the well-known, better known to you than to me, difficulties of divergences in field theories. I point out the desirability of doing some work on it. If mathematicians read Viki Weisskopf's explanations of what physics is all about, it might stimulate a lot of this work.

I spoke of time and space and the other models that are possible, but the situation is even more interesting than that; namely, one can ask, "What is the nature of, what mathematicians call, the primary variable, used in physics?" The Newtonian point of view was essentially that points are the primary variables. The idea of the field came with the role of the continuum of points and it is still incredibly successful.

But think of the new statistics, Fermi–Dirac and Bose–Einstein statistics which deal with points, but with indistinguishable points. And that is really something new again. The ψ-function, the primary variable of quantum theory, is no longer what we consider a geometrical point. A set of points, or a set of sets, is different from the classical collection of points considered in analytic geometry or in most courses in mathematics, even topology.

Now if I may mention something more personal: In 1931 when I was still in Poland, my friend Karol Borsuk and I, considered entities which are finite sets of points and tried to make a topological space of these. That is, consider a finite set of points on an interval and do not distinguish between their order. The element is a set of points and the distance between two sets can be defined in a simple way as described by the mathematician Haudorff. Given a point x of set B, look at the nearest point y in set A and then take the maximum of that minimum with respect to all choices of x. That will be the distance between sets. The reason I mention this is that we wrote a paper, actually it was the first paper I ever published in this country. We sent it to the Bulletin of the American Math Society in 1932. In it we showed that for two dimensions you get a space that you might imagine just as a triangle, for three dimensions you can imagine it as a sort of tetrahedron, but topologically they are the same as the square or a cube. For four dimensions it is still topologically the same. And now surprise! For five or more dimensions one gets a strange sort of manifold topologically different from the n-dimensional cube. At that time I had no idea of the indistinguishability of particles, the statistical mechanics theories of Einstein and Fermi. This I merely mention to show that in mathematics, one considers as a variable not only points but sets. Sets for a space with respect to distance, and as you go to infinitely many dimensions, you get something very interesting in the space of the closed sets. Here again is something that most mathematicians should learn about, namely the statistics of identical particles. Of course these statistics are very unintuitive. Gibbs, himself, found a paradox by considering the entropy of two gases mixing. The entropy increases because of diffusion from one part of the gas to the other, but when the two gases are the same, the formula is different because the diffusion does not count. This was the precursor of the idea of indistinguishability. We don't know which particle is which and the actual formula for the entropy is different. This difference was of fundamental importance many years later. Here again is a very interesting field of work provided one explains to mathematicians what physics deals with, and what is actually important conceptually.

To go on just a little further, I want to describe a very nice new mathematical game on nonlinear problems. Let me tell you the humble and simple beginning of a topic which now has many books and dozens of mathematical papers. The reason I mention it is because no less a physicist than Fermi was involved in it. It was in the very early days of computing and Los Alamos had one of the first working computing machines. I remember discussions in which we looked for good physical problems to study on a computer — a problem in which one could not gauge even the qualitative behavior by calculations as they would take thousands of years. After some time Fermi said, "Well, a nice problem would be to see how a string vibrates if you add to the usual Hooke's law a very small nonlinear force." So this was the problem. We were interested in the rate at which the motion becomes ergodic or mixed up. Fermi, himself, expected that after a while, after many of what would be linear vibrations back and forth, that the string would distort more and more. But nothing of the sort happened. To the great surprise of Fermi, who after all had tremendous experience in problems involving wave motion, it didn't want to do that. Only the first few modes played among themselves. And the higher modes, beginning with say the fifth and up to the 64th, involved all together only one or two percent of total energy. So the motion was very nonergodic and in addition very surprising that after a few thousands of what would have been the normal period, the string came back almost exactly to its initial position. According to a theorem of Poincaré, the return theorem, this should happen sometime, but the time interval for it would be of the age of the universe if you estimate it by phase-space volume. But no, this happened quickly. Our computer work gave rise later to many studies trying to explain this phenomenon. This nonergodic result is also physically observed in some propagations of sound waves in crystal and by now people know that certain nonlinear problems lead to quasi-states or quasi-eigenvalues. In a linear problem, these characteristic states appear just like in the theory of Hilbert space used in quantum theory. But the suspicion was that nonlinear problems led to a general increase of entropy, a mixup of everything. But no, in a large class of problems, even in some problems of classical dynamics, there is no apparent ergodicity under certain conditions.

Fermi said, "Well, it does not say in the Bible that the fundamental equations of physics must be linear." And the idea of quasi-states or what some mathematicians now call solitons, is now being used in some speculations about the more fundamental physical problems, in particular models for elementary particles. Here, again, something that started with very humble, simpleminded calculations has led to interesting mathematical developments and speculations.

This ends my propaganda for mathematicians about some conceptual physics of the type that Viki practices so well and which he also explains so well in his writings.

9

How Aristotle Discovered DNA

Max Delbrück

California Institute of Technology, Pasadena, California

I am wondering how to address you, Viki, on this tremendous occasion, dedicated to commemorate your approaching "graduation from college". Like many of us here assembled, you will have to think of a career to choose after this "graduation". Perhaps the most appropriate form of address would be the way the young Goethe was instructed to address his grandfather, namely, "Erhabener Grosspapa!" That could be translated "Exalted Granddaddy", but the flavor is not quite the same. I'll start out with some comments on Stan Ulam's talk. He invited us to speak up in the discussion to his talk, but I prefer to do it now when I have the floor to myself, so he can't talk back. There are several of his quotes that I want to comment on. One quote from Fermi on some theory that had been confirmed better than he, Fermi, thought the theory had any business of being that good. To anybody that works in biology and is aware of the fact that our brain evolved to help us get along in the cave, it is utterly miraculous and completely incomprehensible that this brain is capable of doing science at the success rate at which it is doing it. This is an aspect that mathematicians and physicists and most scientists tend to ignore. But it is one that is very much in the minds of those who are trying to understand more deeply the nature of our perceptive and cognitive capabilities from the point of view of biology.

Along the same lines, I was interested to hear Ulam refer to the "external world". He didn't really say what the opposite to the external world was, except that at one point he implied that maybe it was the world of mathematics as created by the mathematicians. Now this is a very ambiguous way of speaking. Does the mathematician *create* the mathematical world or does he *discover* the mathematical world? After all, the 2×10^5 theorems per year that are published, no mathematician would admit that these are free creations. I think most mathematicians look upon the world which they discover as a far more real world than the one that Ulam referred to as the external world. The mathematician deals with a "real" world which has a more absolute existence than the "external" world. We know very well that there is an ambiguity about defining an external world, we know that from quantum mechanics. But certainly there is no uncertainty principle of that kind about the mathematical world. No mathematician would admit any uncertainty about, say, the fundamental theorem of arithmetic, the theorem that says that every natural number can be decomposed into prime numbers in only one way. That is a deep theorem that had to be discovered and that is a discovery that refers to a world that is totally real to the mathematician and totally independent of the mathematician, from the point of view of the mathematician. The Godel theorem to which Ulam referred states that even in classical mathematics the system of axioms and procedural rules is limited in the sense that there must exist theorems that are not decidable. That theorem does not detract from the reality of the world. It only makes it, in a way, easier for mathematicians to cope. After all, somebody who's trying very hard to prove or disprove Riemann's Hypothesis may take comfort in the fact that it is not his own stupidity that frustrates him but the fact that the question he is attempting to settle may not be decidable at all. In fact, there may be very many such theorems so he can take comfort in it.

Finally, I want to comment on his quip that anything is hard to predict, especially the future. I think that Ulam is wrong. The future is easy to predict, it is obvious: the future lies ahead! That's the answer. It means *time has a direction* and what Ulam was referring to is not that it's difficult to predict the future but that you cannot get this fact that time has a direction from physics. It is a well-known fact that the laws of physics do not explain one of the most elementary facts, namely, that the evolution of the world has a direction in time, which reflects one of the limitations of physics and of thinking.

Now I come to my talk. I first want to clarify which Aristotle I am talking about. The Aristotle that I am talking about is *not*: the Athenian statesman, the author of graceful forensic speeches; is not the scholar who commented on the *Iliad*; is not the Sicillian rhetorician who wrote a reply to the panegyric of Isocrates; is not the disciple of Aeschines, the Socratic philosopher, surnamed Myth; is not the native of Cyrene who wrote upon the art of poetry; is not the trainer of boys mentioned by Aristoxenos in his *Life of Plato*; is not the obscure grammarian whose handbook on redundancy was still extant at the time of Diogenes Laertius, who lists all the above; and is not, finally, the husband of Jackie Kennedy; but it is the man who was born in the first year of the 99th Olympiad, 384 B.C., at the time when Plato was 44. He enrolled in Plato's academy at the age of 17, when Plato was 60. Like presidents of other colleges, Plato was not around when our Aristotle enrolled. In fact, he was away for two years. That, I think would not be permitted these days. He was on a trip to Sicily. Not until two or three years later did Plato come back. But Aristotle hung around long enough, about 20 years, to see something of Plato and hear something and also read something. Aristotle was exceptional in that he read. In those days, books were very expensive and few people, few students, had the money to buy books. And publication really consisted in giving a lecture at that academy. There was also a book market, you might say, and the Aristotle that I am talking about did write a number of dialogues which were popular explanations of science and philosophy at that time. He published them in the Scientific Athenian. All of these are lost. But he stayed around for 20 years and then went traveling, from the age of about 36, and then came back to Athens and founded his own school, after he had not been made successor of Plato as president of that college. He founded his own school in some other part of town, at a convenient place to walk around with his students. He spoke with a lisp. "His calves were slender, his eyes small and he was conspicuous in his attire."

Aristotle was a biologist who wrote a number of books on biology: *The Historia Animalium*; *De Partibus Animalium*; *De Motu Animalium*; *De Incessu Animalium*; *De Generatione Animalium*. Five books, some larger, some smaller, full of an enormous amount of detail, and even more full of a great deal of speculation. A great many of the details are amazing and some of them that were laughed at for 2,000 years actually have turned out to be correct on more recent scrutiny. Many of them, of course, not. He did not have a microscope. But he did do a lot of dissecting, or rather he watched fisherman and other friends of his dissecting animals, and he also studied plants at great length, and was aware of these biological objects from his earliest youth, coming from a physician's family from both his father's and his mother's side. This not only gave him familiarity with the facts of life, not only the external but the internal facts of life, but it also gave him sufficient money to buy books. His nickname was "the reader" because his financial status allowed him to collect papyrus rolls and establish a library containing quite a number of them. So he had this exposure to "evolving things", this imprinting of "end" or "purpose" from being most intimately familiar with medicine from childhood on and then with many animals and plants from direct study and extensive reading.

The most pervasive concept of his philosophy and of his writing is the fact that these animals, these creatures, these objects develop apparently according to a *preimposed plan*. Let me first read you a passage from one of his books, *De Partibus Animalium*. It says, "*of the products of nature, some are eternal, not subject to generation and corruption.*" "For 'generation' we might today maybe better use the word "development". I would like to say, parenthetically, that it is very ambiguous how to translate Aristotle, how to translate anybody, but specifically Aristotle, for a number of reasons. One of the main reasons is that our dictionaries of the Greek language were made at a time at which our understanding of what he had to say was beclouded, as I will show. "*So some are eternal, those are the celestial objects, and grow and perish. Of the former, grand and divine as they are, we have less insight since they offer few aspects for our perception.*" That was true in those days before telescopes and radio astronomy, and so on. All that could be done was map the fixed stars and map the courses of the planets. That was a business that had been going on and I am told that between the time that Plato was a young man and between the time that Plato was an old man, it had become clear that the erratic movements of the planets were not totally erratic but that there was a system behind it. So at the time Aristotle came around, he had been informed, by reading the appropriate papers, that the planetary motions presumably implied a cosmological order. As he rightly says, there really isn't very much you can do about it. You just note the courses of the planets and then you try to make a theory about it. "*From these scanty data we can explore what we care to know about them,*" and obviously he did not care. "*In contrast, for the perishable things, plants and animals, we are given a great wealth of information since they are close at hand. If one makes the effort, much can be learned about each kind. Both sciences have their own charm. Even though our understanding of the eternal things is more limited they fascinate us more than the things of our own world because of their grandeur. Just as our imagination gets more excited by even a glimpse of a beloved person than by the close observation of many other, and even more important things. However, the perishable things are to be preferred as objects of science because or the wealth of knowledge we can gather about them.*" This was before the days when there were 900 journals of biology. "*I will speak about the nature of animals and, to the best of my ability, not overlook anything, may it seem large or small. Also, with those less appealing creatures that many of us are shy of looking at, nature grants indescribable pleasures to those of the scientific bent by revealing her creative power to their scientific scrutiny.*" In translating I use the word "Nature" and not "God" or anything else, leaving it open what the nature of this nature is, which has such creative power. "*Indeed, it would be absurd were we to take delight in artistic reproductions, admiring the craft of the artist — as we do with paintings and sculptures — and not take delight in the original creations of nature, especially when we can achieve some measure of understanding of their structure. Therefore, one should not childishly recoil from the study of lower animals. All creations of nature are miraculous. When strangers were visiting Heraclitus and found him warming himself by the kitchen stove, they hesitated to enter. He encouraged them to approach saying, "The Gods are here, too." Just so one should approach the study of any animal or plant with reverence, in the certainty that any of them are natural and beautiful. I say 'beautiful' because in the works of nature and precisely in them there is always a plan and nothing accidental.*" (That may be an exaggeration.) "*The full realization of the plan, however, that for which a thing exists and towards which it has developed, is its essential beauty. Also, one should have clearly in mind that one is not studying an organ or vessel for its own sake but for the sake of the functional whole. One deals with a house, not with bricks, loam or wood. Thus, the natural scientist deals with the functional whole, not with its parts, which, as separate entities, have no existence.*" We can feel that he is defensive about his biology and he is defensive in the environment of the Platonic tradition which had very little relation to the world of biology and to this primary impression that we get here, that things carry within them, or have inscribed in them, a plan according to which they develop. In another passage in De Generatione Animalium, he says, "Since it is impossible that creatures should be eternal, these things are generated are not eternal as individuals (though the essence is in the individual), but as a species." This formulation is a dig at Plato who would have said that there is an idea of the dog is the real thing that carries the plan within it.

Anybody who is familiar with today's physics and biology and who reads Aristotle's writings in these two fields must be struck by the aptness of many of his biological concepts in contrast to his physics and his cosmology. Physics was practically non-existent and also there was very little interest along that line. Nobody would deny that Aristotle's physics was pretty much of a catastrophe while his biology abounds in aggressive speculative analysis, in vast observations on morphology, anatomy, systematics and most importantly, on embryology and development. Aristotle *does* consider it remarkable and does consider it a fundamental aspect of nature that "*human beings beget human beings and do not beget rabbits or an ear of corn.*" * What strikes the modern reader most forcibly is Aristotle's insistence that in the generation of animals the male contributes in the semen a form principle, not a mini-man. Leaving aside his male chauvinism which made him think that the plan was contributed by the male only, leaving that aside, the question was, does the male do so by contributing in his semen extracts from all his organs, from all parts of his body? This had been Hippocrates' theory. Aristotle points out, first of all, that the resemblance of children to parents is no proof of part-to-part representation because the resemblance is also found in voice in ways of moving and behavior patterns. Secondly, that men generate progeny before they have certain parts such as beards or gray hair. And similarly with plants. Thirdly, inheritance may skip generations as in the case of "*a woman in Elis who had intercourse with an Ethiopian. Her daughter was not dark but the daughter's son was.*" Finally, about the inheritance of sex, he argued that since the semen which determines the form, can determine female children, it clearly cannot do so by being a secretion in a man from female genitals because the man does not have female genitals. So, from this it is clear that the semen does not consist of contributions from all parts of the body of the male and the female's contribution is quite different from the male's. The male contributes the plan of development and the female the substrate. For this reason the female does

---

*This is a quotation from a short lecture on molecular genetics by Joe Herscovitz, entitled, "The Double Talking Helix Blues", a five minute lecture (with guitar on molecular genetics, available as a phonograph record published by The Vertebral Disk, 913 S. Claremont, Chicago, Illinois 60612.

not produce offspring by herself since the form principle is missing. This was his theory of impossibility of parthenogenesis, although he was aware of some cases where parthenogenesis does occur. For those other animals he had to have a special theory. In normal animals the male has to come and contribute the form principle, the plan. The form principle is likened to a carpenter. The carpenter is a moving force which changes the substrate. But the moving force is not materially contained, in the finished product. The semen contributes nothing to the material body of the embryo, it only communicates its program of development. This capability is that which acts and creates.

The material which receives this instruction and is shaped by it is the undischarged residue of the menstrual fluid (slightly in error, but that's understandable). The creature is produced from them: the form principle in the semen and matter coming from the female. They, together, produce the creature. "*It is produced like a bed come into being from the carpenter and the wood.*" "*The male contributes the principle of development, the female the material.*" These are all quotes from various Aristotelian writings. "*The male emits semen in some animals and where he does it does not become part of the embryo,*" he thought. Remember, he did not have a microscope. Quite a few quotations in a similar vein could be added. Put into modern language what all these quotations say is this: After fertilization, it is read out in a pre-programmed way. The readout organizes the matter upon which it acts but it does not organize the stored information, which is not properly speaking part of the finished product. In paraphrasing Aristotle this way I do not think I have changed a thing of what he said — I've changed just the wording, the way we would word it today. Therefore, I think that the committee in Stockholm which has the unenviable task each year of pointing out the most creative scientists, if they had the liberty of giving awards posthumously, I think they should consider Aristotle for the discovery of the principle implied in DNA.

It is my contention that Aristotle's principle of the unmoved mover originated with his biological studies and that it was grafted from here first onto physics, then onto astronomy, and finally onto cosmological theology. I should like to suggest, furthermore, that the reason for the lack of appreciation among scientists for Aristotle's scheme lies in our having been blinded for 300 years by the Newtonian view of the world. So much so, that anybody who held that the mover had to be in contact with the moved and talked about an unmoved mover, anybody who did that collided head on with Newton's dictum "action equals reaction." Any statement in conflict with this axiom of Newtonian dynamics could only appear to be muddled nonsense, a left-over from a benighted, prescientific past. And yet, unmoved mover perfectly describes DNA. DNA acts, creates form in development, and it does not change in the process.

I want to go one step further and consider the fact that the re-entry of Aristotle into western thought occurred through scholastic, Christian theology in the 12th and 13th century. The history of Aristotle's manuscripts is a very crazy one. For 300 years, they were stored, they were carried from Athens to Asia Minor and back to Athens and then about 50 B.C. they were bought up by a bibliophile and brought to Rome and were published there. In the Middle Ages, they were forgotten in the West and then, at the time of scholastic studies, they were appropriated by the theologians. And through the theologians Aristotle came back into the West through the application of his schemes to theology, to logic, his own logic, and so on. This is the tradition in which Aristotle has come through to the later centuries. So by the irony of history, the vast historical impact of Aristotle on western thought is the result of an almost accidental appropriation of the most secondary and misguided aspects of Aristotle's speculations.

It is due to this bizarre twist that we are encumbered today with a total barrier of understanding between the scientists and the theologians, from St. Thomas Aquinas to today. Catholic, Protestant, and LSD mystic alike. So I want to propose that a new look at Aristotle, the biologist, may yet lead to a clearer understanding of the concepts of purpose and of truth and of their relation, and it may lead perhaps to even something better than mere coexistence between the natural scientists and our colleagues from the other faculties. This mere coexistence which prevails now is reflected in the fact that these several faculties really have very different notions of reality and of truth. I referred earlier in making comment on Ulam's talk to the fact that the mathematicians live in their world, which to them is more real than what is commonly referred to as the physical world. And that the concept of truth in mathematics has a far different meaning than it has in physics. When we come to biology and psychobiology and to the questions of language and meaning, you will find that the ground shifts again tremendously. And this shifting ground and this operating with very distinct concepts can be described, but it cannot at the present time be reconciled. And I think it cannot be reconciled in large part because our language and our concepts fetter us. I think it is very difficult to unthink schemes which we have found convenient to use, modes of talk, modes of intellectual intercourse which we have developed. I do not think that probing into the early origins of these concepts and these conceptual relations is idle, but that it is, in fact, very necessary.

10

Is Negotiated Arms Control Possible?

W. K. H. Panofsky

Stanford University, Stanford, California

I had a very difficult time deciding on the topic of this talk, since Viki's interests cover such a broad range of activities with which I am also concerned. You can hear next week about the recent exciting work with the SLAC storage rings, a description of the design principles of such rings, and their future promise for new physics through Professor Richter's Loeb Lectures at Harvard. Talking about inelastic lepton scattering during an M.I.T. conference would be bringing coals to Newcastle, since the local M.I.T. physicists are primary agents in these experiments. Broad problems in high energy physics policy, for instance such questions as the relation between University users and the large laboratories, are matters of current concern to Viki and his friends in high energy physics, but I doubt whether many would sit still for a one-hour talk on that subject. I would therefore like to use the opportunity to express some personal views on certain current issues in arms control, since I know that there exists a wide spectrum of involvement and also opinion on this subject in the local community.

Many in this audience might ask why the question implied in the title of this talk needs to be asked at all. After all, during the past 15 years there has been a large number of negotiated agreements, starting with the Antartic Treaty in 1959 and ending with the Moscow Treaty signed this year at the Summit Meeting. The next two figures give the total list of such agreements.*

---

*Tables 1 and 2 are taken from Alva Myrdal's article on disarmament in the October 1974 *Scientific American*.

Yet most students of the subject will agree that, however impressive these achievements might appear, they have been largely peripheral to the main thrust of the arms race; in particular, the competition in the strategic nuclear arms race between the Soviet Union and the United States may have been deflected, but it has surely not been arrested. Most of the recent arms control agreements have been justly criticized as limiting only those activities which the participants in negotiations did not wish to undertake anyhow, or else they affected only those weapons which could easily be replaced by other equally lethal alternatives. In fact, when looking at the United States' military effort, it is difficult to discern much, if any, restraining effect of negotiated arms control agreements. Most would agree that the constraint of negotiated arms control has been negligible relative to the normal limiting effect which proceeds through the budgetary process, and most financial restraint has

Table 1.   Multilateral arms agreements.

Signed	In force			Number of parties
1959	1961	Antarctic Treaty	Prohibits all Military Activity in Antartic Area	17
1953	1963	Partial Nuclear Test Ban Treaty	Prohibits Nuclear Explosions in the Atmosphere in Outer Space and Under Water	106
1967	1967	Outer Space Treaty	Prohibits all Military Activity in Outer Space Including the Moon and Other Celestial Bodies	71
1967	1967	Treaty of Tlatelolco	Prohibits Nuclear Weapons in Latin America	18
1968	1970	Non-proliferation Treaty	Prohibits Acquisition of Nuclear Weapons by Non-Nuclear Nations	82
1971	1972	Sea-Bed Treaty	Prohibits Emplacement of Nuclear Weapons and Other Weapons of Mass Destruction on Ocean Floor or Subsoil Thereof	52
1972		Biological Weapons Convention	Prohibits Development, Production and Stockpiling of Bacteriological and Toxin Weapons and Requires Destruction of Existing Biological Weapons	31

Table 2.   Bilateral arms agreements between the U.S. and the U.S.S.R..

Signed	In force		
1963	1963	"Hot Line"	Establishes Direct Radio and Telegraph Communications Between U.S. and U.S.S.R. for Use in Emergency
1971	1971	"Hot Line" Modernization Agreement	Increases Reliability of Original "Hot Line" System by Adding Two Satellite-Communications Circuits
1971	1971	Nuclear Accidents Agreement	Institutes Various Measures to Reduce Risk of Accidental Nuclear War Between U.S. and U.S.S.R.
1972	1972	High-Seas Agreement	Provides for Measures to Help Prevent Dangerous Incidents on or over the High Seas Involving Ships and Aircraft of both parties
1972	1972	Salt I ABM Treaty	Limits Deployment of Anti-Ballistic-Missile Systems to Two Sites in Each Country
1972	1972	Salt I Interim Offensive Arms Agreement	Provides for Five-Year Freeze on Aggregate Number of Fixed Land-Based intercontinental Ballistic Missiles (ICBM's) and Submarine-Lanched Ballistic Missiles (SLBM's) on each side
1973	1973	Protocol to High-Seas Agreement	Prohibits Simulated Attacks by Ships and Aircraft of each party aimed at Nonmilitary Ships of Other Party
1973	1973	Nuclear War Prevention Agreement	Institutes various Measures to Help Avert Outbreak of Nuclear War in Crisis Situations
1974		Salt II ABM Treaty	Limits Deployment of Anti-Ballistic-Missile Systems to One Site in Each Country
1974		Salt II Threshold Nuclear Test Ban Treaty	Prohibits Underground Tests of Nuclear Weapons with Ex-plosive Yields Greater than 150 Kilotons
1974		Salt II Interim Offensive Arms Agreement	Commits Both Parties to Negotiate Extension of Salt I Interim Offensive Arms Agreement Through 1985

not been a very powerful instrument in dealing with the problem, as many of the now well-known indices of the arms race indicate.

I do not have to tell this audience about the implied horror of the numbers which delineate the magnitude of the problem. The world spends more on armament than on education. The rate of weapons spending has been considerably greater in the lesser developed countries than in the developed countries. Military sales of the developed to the lesser developed countries have risen dramatically in recent years. The inventories of nuclear weapons of those nations which have joined the nuclear club correspond to tens of tons of TNT equivalent for each inhabitant of this globe. The physical, biological and ecological side effects of large scale, or even medium scale, nuclear war are poorly understood; it has recently been pointed out by the Director of the Arms Control and Disarmament Agency that those side effects which are known have largely been discovered by accident. Yet in the face of this ignorance nuclear weapons continue to be used as diplomatic instruments in attempting to reach favorable international accommodation, although no use "in anger" has occurred since Hiroshima and Hagasaki. There are those who, in spite of all these facts, still claim that there is no arms race. They point to the undeniable fact that military spending in the U.S. has lately been a decreasing fraction of the Gross National Product and that, depending on bookkeeping practices, such spending has also been a decreasing fraction of Federal expenditures. Yet there are two salient facts contradicting that optimistic conclusion:

1. Government personnel and direct purchases by the Federal Government continue to be about three-fourths military. This accounts for the deep involvement of the very fiber of government with military goals.
2. The arms race cannot be measured fairly by the *rate* of expenditure. Rather the problem, in particular in the area of strategic nuclear weapons, is the *accumulation* of stockpiles; they continue to grow even when the rate of acquisition may at times appear to be slowed.

Possibly the most disturbing evolution is the use of nuclear weapons inventories for political rather than military purposes. Many will respond to this comment by saying, "So what else is new?" Military force has always been viewed as an instrument of political leverage rather than an end in itself. After all, it is argued that the purpose of a military establishment has always been to further what are ultimately political or economic goals. However during the last decade the nuclear arms race has added a new dimension to this problem: In the past, military power was considered to be "usable", and therefore its magnitude bore some relationship to the social or political aims in sight. Now most statesmen conclude that nuclear weapons can no longer be "rationally used"; there are no longer any believable or predictable specific scenarios of war for which they can be specifically designed. **However, it is just this absence of projected "rational use" of nuclear weapons which has also removed any rational limit on their number and character.**

This situation has been reflected in the constantly fluctuating rationales for the procurement of nuclear weapons which have been introduced by successive generations of Secretaries of Defense, or even the changes in strategic policy which have been made during the tenure of a given Secretary. Moreover, just the very absence of any real logic for the procurement of a specific level of nuclear armaments has removed any rational motive for unilateral restraint, and the magnitude of nuclear armament has become largely a symbol of such unquantifiable aims as "national resolve", "superiority", "national prestige", etc. Changing national strategic policies, including the much discussed recent Schlesinger doctrine, and changing nuclear armaments cannot change the basic physical fact that the populations of the Soviet Union and the United States — and to a lesser extent the population of other countries — are hostages in the sense that they can be destroyed at the will of an opponent. Political use of the existing weapons inventories tends to proceed in apparent oblivion of the physical realities these weapons represent.

Thus, despite the superficially impressive list of negotiated arms control agreements, a very depressing picture remains. Another factor deepens the gloom further and has led many to doubt the utility of pursuing negotiated arms control at all in the future: this is the fact that the existence of arms control negotiations has frequently resulted in an actual acceleration of the arms race in some of its aspects. These adverse aspects fall into a number of different categories, which I will now discuss in further detail.

First there is the *"bargaining chip"* argument, by which national leaders maintain that while arms control negotiations are in progress, military procurements must be accelerated beyond a rate that might otherwise be considered prudent. The claimed objective is to exert pressure on the opponent through the potential threat of new armaments which one side would be willing to give away if appropriate counter concessions were made by the other. Let me remind you that the Safeguard ABM deployment was passed by the Congress in the face of solid technical criticism only after the Administration maintained it was needed as a "bargaining chip" at SALT.

The second escalatory effect is the so-called "*safeguards*" philosophy; for instance, when President Kennedy submitted the 1963 Limited Nuclear Test Ban Treaty for ratification by the Senate, he had to pay for the needed non-opposition of the military establishment by agreeing to a set of four safeguards which included, among other factors, the commitment to carry out underground nuclear testing at a pace described as "comprehensive, aggressive, and continuing". As a result, the rate of nuclear testing, although driven underground by the 1963 Treaty, was in fact accelerated.

A third escalating effect is what I might call the "*get in under the wire*" argument: if an arms control agreement is imminent, then military parties on all sides exert pressure to accelerate work before the deadline. A typical example is the recent supplementary budget request of the Atomic Energy Commission to accelerate nuclear testing above the 150 kiloton limit which might be imposed by March 31, 1976, provided the agreements negotiated in Moscow this spring are ratified by the Senate.

Another detrimental effect of negotiated arms control steins from the excessive *emphasis on "nuclear numerology"* generated by the negotiation and ratification process itself. Let me give an example: The SALT I Interim Agreement resulted in a limitation in numbers of missile launchers which appeared to favor the Soviet Union; in contrast, in terms of deliverable re-entry vehicles which were not controlled by the agreement, the U.S. remains far ahead. The resultant debate in the Congress and elsewhere pitted Hawks and Doves against one another in arguing "who is ahead" in view of this complex situation. This debate gave these numerical questions undue public and political importance, while in fact the potential strategic utility of the missile forces of both countries would depend a great deal more on such characteristics as accuracy and reliability and on the quality of command control associated with these weapons than on their actual detailed numbers.

Finally, an adverse effect inherent in some arms control negotiations is the *excessive emphasis generated on peaceful applications* of military technology. For instance, in order to erase the stigma associated with the awesome destruction made possible by nuclear weapons, President Eisenhower introduced his "Atoms for Peace" program, which was intended to emphasize the benefits of nuclear reactor technology. Subsequently, reactor technology and reactors have been used and are still being used as components of foreign aid programs. I would dare say that the inclusion of reactors in such programs has only partially been motivated by the true economic value which either research or power reactors would confer on the recipients; a primary reason is the prestige value thus generated. As the recent Indian nuclear explosion testifies, the international safeguards arrangements designed to prevent diversion of reactor fuels, which I will not discuss here in detail, have not always been adequate to prevent diversion of the nuclear fuel for producing nuclear explosions. The most recent case in point is what I consider to be excessive emphasis on the peaceful uses of nuclear explosions (PNE's). In accommodation to a Soviet request, PNE's were exempted from the recent Moscow Threshold Test Ban Treaty; also in the Non-proliferation Treaty a clause was introduced which obligates the nuclear "have" nations to cooperate with "have not" nations in regard to peaceful applications, provided such a technology is usefully developed. In essence, these examples signify that in past arms control negotiations the participants were unwilling to face the issue that, at times, in order to bring the technological arms race under control, it may also be necessary to forego conjectured peaceful benefits from that technology. In fact, the tendency in the process of negotiation has been to give these beneficial peaceful applications undue merit and a prominence greatly exceeding the economic value which realistically might accrue.

The preceding summary of the limited scope of the arms control agreements as negotiated to date, and the escalatory effects experienced collateral to these negotiations, naturally make many critics conclude that further pursuit of negotiated arms control is, to use the modern jargon, counterproductive. But what are the alternatives? One option is to forget about arms control entirely and to let the political and budgetary processes of each country determine the future fate of the armaments of all nations — clearly a very dangerous course; nevertheless many voices are heard that an arms race "can be won" due to the economic strength of this country. I question not only the morality of this course but also its economic truth; it may actually be true that the sensitivity of the Western economies to the continued arms expenditures is higher in this inflationary period than that of controlled economies. To quote an October 1 New York Times dispatch: "In the Pentagon view, the inflation situation is so serious that some defense planners are now gloomily talking about how the United States is being forced into what they term 'a slow process of unilateral disarmament'." In terms of economic impact, a dollar spent on military expenditures tends to be more inflationary than other forms of government spending which fill an unsatisfied need.

Another alternative to negotiated arms control is the exercise of "arms control by mutual example", in other words by unilateral restraint, publicly advertised, which is hopefully then followed by reciprocal action on the other side. The problem is that while such a mechanism for arms control may be effective at times, it would in general tend to be fragile if extended over any significant length of time. It is almost unimaginable that if, following a substantial period of unilateral restraint on military spending — for example by the U.S. — the Soviet Union did not reciprocate, it would be possible for the U.S. to continue restraint under increasing political pressures, even if it were militarily safe to do so.

My conclusion is therefore that the question raised with increasing frequency — Should arms control be negotiated or be generated through a policy of mutual restraint? — deserves not an either/or answer, but rather a combination of these two efforts must continue. It is such a combination which can minimize the negative side effects of negotiated arms control cited above.

Let me illustrate my conclusion that a *combination* of negotiated arms control and mutual restraint is both necessary and desirable if there is any hope of arresting the nuclear arms race by reviewing briefly the achievements and deficiencies of both the SALT I Treaty on ABM Limitations, signed in Moscow in 1972 (and its annexed Protocol signed July 3, 1974) and also the Threshold Test Ban Agreement signed on July 3, 1974. I hope to illustrate thereby a "good" treaty resulting in an arms limitation step which could not have been achieved solely through mutual restraint, while the Threshold Test Ban is an example of what I consider to be a "bad" treaty whose deficiencies may well exceed its gains.

The 1972 ABM Treaty was an important step in arms control. By limiting deployment of ABM systems to what are in essence insignificant levels for strategic purposes, it implements its basic policy declaration: "Each party undertakes not to deploy ABM systems for the defense of the territory of its country ..." This declaration in fact codifies the mutual hostage condition of the population of each, country; it is this conclusion which is based on physical fact, not strategic doctrine, as long as strategic arms are not drastically reduced.

It is true that there has been "Restraint" on both sides of the Iron Curtain in building ABM systems, even without a treaty. However this restraint was not so much a result of deliberate arms control as it was a consequence of the technical difficulty of building an ABM system which has any military promise at all. The Soviets have carried out a vigorous ABM research and development test activity at Sary Shagan near Lake Balkash; they started to build a defense around Leningrad more than a decade ago only to take it down again. They then constructed the presently existing 64-interceptor Moscow defense which is recognized by all to be of very limited value in decreasing the impact of any but the most limited ICBM attack. Nevertheless, in view of the still continuing vigorous development and test activity, there is little question that **additional Soviet ABM systems would have evolved in time were it not for this treaty.** On the U.S. side extensive research and development activity has also been going on, including the publicized successful intercepts on Kwajalein Atoll. A decision to deploy an ABM system was deferred under Secretary McNamara on a year-by-year basis until in 1967 he proposed to go ahead with a thin area defense called the Sentinel System which was to offer some minimal protection against accidental or non-Soviet attack; at that time this meant attack from China which at that time was officially considered to be less "rational" in potential action than the Soviet Union.

The reason for this hesitant decision stemmed both from the recognition of the limited ability of an ABM system to offer significant protection against a massive missile attack and also from the understanding of the escalatory nature of ABM deployment to the arms race. Yet political pressures forced deployment. As is well known, the mission of Sentinel was redefined under the Nixon Administration to become primarily a defense of U.S. missile silos; the presently deployed, but soon to be mothballed, Safeguard installation near Grand Forks is the result. It is clear from this history that **without the ABM treaty the U.S. political process would not have resisted further expansion of this system** or the deployment of a follow-on hard-site defense around missile silos. Self restraint on any one side in deploying ABM would not have continued once expanded deployments were seen on the other, or once it was conjectured from observation of the opponents' R&D program that future expanded ABM deployments were in the offing.

The Protocol of July 3, 1974 has amended the 1972 ABM treaty by removing a defect of the earlier treaty which, was important in principle, although not important in practice as applied to the actual situation. The 1970 ABM Treaty deserved criticism in detail in that it was in effect "arms control upwards". Both sides kept the one ABM site they had, but in addition were permitted to add another site corresponding to what the other side had built or had under construction.

The 1974 Protocol recognizes the existing asymmetry between the parties by restricting each side to the one ABM site actually constructed; formal symmetry is preserved by permitting each side to destroy its existing ABM site and convert it to the type of deployment that exists on the other side. Needless to say, there is little expectation that such a move will actually be made. Although the 1972 Treaty restricted any future ABM deployment to levels which are negligible from a military (but not a budgetary!) point of view, the removal of the principle of "arms control by permitting one side to buy what the other side already has" is highly laudable. In fact what the ABM 1974 Protocol actually establishes is a new and valuable principle in arms control which is similar to a successful practice of parents dividing a cake for children: the principle is to let one kid cut the cake and the other one choose which piece to eat. The 1974 Protocol essentially freezes the existing systems which are asymmetric in character; however each side is given the option to convert to the deployment pattern of the opposite side if they think this to be more valuable. The precedent established by this provision is very important. Many U.S. critics of the SALT I agreement and of the existing strategic nuclear weapons programs point to inferiority or imminent inferiority of U.S. forces relative to the Soviets by selectively considering those systems where the Soviets are ahead; yet when faced with the question, as Secretary Schlesinger recently was, "Would you trade the Russian system for our own?" the answer turns out to be a resounding "No". I hope that this type of principle, which permits mutually acceptable agreed arms control patterns even in the face of a very asymmetrical strategic situation, will have application in the future.

While the ABM Treaty of 1972, as improved by the supplementary Protocol of 1974, illustrates the positive value of a negotiated agreement, it also demonstrates that much of what has been accomplished can be negated if restraint is not exercised at the same time. In principle one of the consequences of the ABM Treaty and Protocol *should* have been to decrease the pressure for increased *offensive* nuclear missiles. The reason is simple: The absence of an effective ABM, as codified by the ABM Treaty, enhances the deterrent value of each reentry vehicle carried by U.S. and Soviet missiles, since potential attrition through expanded ABM defenses is no longer of concern. Unfortunately, this rather obvious benefit has not been used by either nation: As is well known the Soviets have aggressively continued an intensive missile test program in order to remedy their technological inferiority in MIRV weapons, while the Americans have embarked on a new series of research and development programs using the "bargaining chip" argument as a basis for the maintenance of just that technological superiority which the Soviets are trying to erase by accelerated testing.

Let me now turn to some critical comments about the Threshold Test Ban Treaty, also signed July 3, 1974, but not as yet submitted for ratification to the Senate. It is well known that because of the debate about the difficulty of distinguishing earthquakes from nuclear explosion it proved impossible in spite of the long series of negotiations from 1958 through 1963 to arrive at a comprehensive nuclear test ban treaty. Rather, the Limited Test Ban Treaty of 1963 was concluded which prohibited testing in the atmosphere, in outer space and under water, but permitted testing underground. However, both that treaty and the 1970 Nuclear Non-Proliferation Treaty committed the signatories to "best efforts" to work towards a treaty banning tests in all media. The Threshold Ban Treaty, signed in July in Moscow, now commits the USA and USSR to restricting underground nuclear explosions for military purposes to values of less than 150 kilotons and located at established test sites. However it specifically exempts peaceful uses of nuclear explosions from this restriction but agrees that a separate agreement safeguarding abuse of such "peaceful" tests should be negotiated as soon as possible. As the treaty is now written there is some question whether it will ever come into force. In principle the Administration could submit the treaty to the Senate for ratification even without an agreed peaceful uses annex. However, and I believe wisely, the Administration has chosen not to do so.

Why is it that so many arms controllers and Doves are opposed to this agreement as drafted in Moscow — after all, is not the banning of underground nuclear explosion above 150 kt a small but definite step forward in arms control. Taken in isolation the answer is, yes, of course it is a step forward, although people familiar with the military applications of nuclear weapons above 150 kt will agree that the limitations on military devices through this treaty are almost negligible; moreover there is a large delay — to March 31, 1976 — built into the treaty.

However, the treaty raises many questions beyond the limitations actually imposed. The official position which the United States has taken throughout the past years is that it would be willing to agree to a comprehensive test ban of nuclear tests in all media provided it can be "adequately verified". During the Kennedy Administration, "adequately verified" meant on-site inspections of the suspected seismic events in order to check locally whether they correspond to nuclear explosions or natural earthquakes. Some of you may recall that there developed a numerical hassle whether the right number of such inspections should be 3, 7, or 0; this controversy, while never having borne much relationship to what such inspections could actually accomplish, was never settled. During the last decade seismic research has progressed to the extent that a number of facts are clear:

1. The limit below which seismic events cannot be detected and the limit below which they cannot be identified whether they are earthquakes or explosions have been lowered; of particular significance is the fact that those two limits have moved closer together. As a result it is now clear that the *incremental* value which onsite inspections would offer over other means of documenting possible evasions is all but totally negligible.

2. It is still true and will remain true that below *some* seismic magnitude or below some size of nuclear explosion there are no physical means of examining whether the opponent has cheated. Yet it is generally agreed that, using seismological techniques alone, evasions can be established with only very few, if any, events remaining unidentifiable below a Richter magnitude of 4.50, which corresponds to an explosion of around 10 kt if carried out at the Soviet nuclear test site of Semipalatinsk.

There is therefore no escaping the conclusion that the "threshold" of 150 kt was not set by considerations of seismic detection but rather to accommodate those on both sides of the Iron Curtain who wished to continue nuclear testing with not too significant impairment, while at the same time showing to the world that some kind of agreement had been reached in Moscow at the Nixon–Brezhnev Summit.

The defects of this agreement signal to the world a disregard by the Soviet Union and by the U.S. of the obligations taken in the Limited Nuclear Test Ban Treaty as well as the Non-Proliferation Treaty that both nations were going to work in good faith towards the conclusion of a comprehensive nuclear test ban treaty. This apparently cynical approach, therefore, may well turn out to be a further setback in the battle against proliferation of nuclear weapons. One should also note that both the Limited Test Ban of 1963 and the Non-Proliferation Treaty of 1967 were signed by many other nations. However, the Moscow Threshold Treaty is bilateral between the superpowers only, and it would be unwise if not in fact illegal to throw the treaty open for signature to other nations not now having nuclear weapons: The Moscow Threshold Treaty might well be considered by non-nuclear nations as a *license* to test nuclear devices up to 150 kilotons rather than as a *constraint* to refrain from developing nuclear weapons on their own.

The disregard of the dangers of further proliferation of nuclear weapons implied by the Moscow Treaty is emphasized even more strongly by the peaceful nuclear explosions loophole built into the Treaty. As mentioned before it is specified that PNE's shall be unrestricted until a separate agreement to govern their control has been negotiated, and this shall be done "as soon as possible". This clause was negotiated in clear disregard of earlier studies which have concluded that it is very difficult, if not impossible, to prevent abuse of "peaceful" nuclear explosions for military purposes; apparently the implication of the Treaty is that "Science will find a way somehow" to translate this political agreement into a physical possibility.

What is the situation concerning the promise of peaceful nuclear explosions? In the U.S. Project Plowshare has been conducted by the Atomic Energy Commission to develop such peaceful applications of nuclear explosives; however the enthusiasm and accompanying budgets have drastically declined in the ensuing years as the economic and environmental deficiencies of PNE's have become more widely recognized. Interest in excavation projects has essentially disappeared, while projects such as gas recovery and shale extraction continue to be discussed and studied, although test explosions have been specifically excluded from next year's AEC budget. There is again discussion of heat recovery (both geothermal and heat from the explosion itself) resulting from large underground detonations, but the numbers going with such a project are outrageous — one would roughly require one hundred 100 kt explosions per year in order to increase the nation's present electrical energy output by 10%. Actually the numbers which go with the other applications are not much better even if the technical and economic viability of the methods has been demonstrated. For instance, approximately 1,000 explosions would be required to increase the nation's natural gas supply by 5%, and the use of nuclear explosions to generate underground storage caverns is most unattractive once it is recognized that the site of such storage units would have to be near centers of population. There was a flurry of excitement starting about a decade ago when alternate routes and methods of excavation for a new canal were being explored to parallel the present Panama Canal. The initial analyses carried out under the leadership of the nuclear weapons laboratories identified a substantial economic advantage for a canal excavated by nuclear means; however when a special commission to investigate the question was appointed and charged with taking all facts into account, this economic advantage reversed once the impact of all the auxiliary problems, including the need for resettling populations, was recognized.

On the Soviet side interest in peaceful uses of nuclear explosions appears strong, although quite divergent views are being expressed publicly by Soviet scientists; I note that this is quite unusual in such a sensitive matter. A leading project on which Soviet attention has been focused is the Kama-Pechora Canal; this is a project which would divert waters from the sources of the Kama River, which empties into the Arctic, to the sources of the Pechora which in turn runs into the Volga and then into the Caspian Sea. The object is to arrest the process of drying up of the Caspian Sea, which has been in progress for a long time, and would also increase the total irrigation capacity of the Volga Basin. It is clear that several hundred devices above the 150 kt threshold would have to be exploded for this project, and various estimates of the economic gains of using nuclear explosions for this purpose have been circulated. In view of the Panama Canal history I have very little confidence that any of the estimates is reliable. The Soviets have done some other unusual things with nuclear explosions; for instance, they have used them to extinguish oil well fires. Needless to say, other methods involving drilling a shaft close to the well in question are available for that purpose; in some respects the use of nuclear explosions here reflects inaccurate drilling techniques rather than a technological triumph.

No identified peaceful uses of nuclear explosions accomplish goals which cannot be reached by conventional means; whether any of these projects do or do not have superior economic merit is a very open question once all factors are taken into account. Above all, it is clear that all such projects, even if they do offer economic advantage, would require a very large number of explosions in order to have a significant impact on the problems to be solved. It is this latter factor which tends to be forgotten by peaceful uses enthusiasts, and it is these large numbers which make the evasion opportunities of a peaceful uses program towards military goals so plentiful. Swords to be forged into plowshares can be passed through loopholes (see figure drawn by a daughter at age 13). At Moscow, Secretary Kissinger revealed that in the preliminary discussions the Soviets appeared not to object to the presence of inspectors at their peaceful nuclear explosion sites. It would indeed be remarkable if after the earlier test ban history in which the Soviets and American negotiations broke down for failure to narrow a gap between three and seven inspections per year now suddenly highly intrusive inspection of hundreds of events would be acceptable.

Quite apart from this overall question the problem remains **what the inspectors would inspect** if they were permitted to be on site of a peaceful nuclear explosion. First of all, is there a basic distinction between a peaceful and military explosion? Each "peaceful" application has some considerable degree of overlap with military applications.

Even if a set of useful criteria could be developed which would at least limit the military usefulness of a "peaceful" test, the problem of verifying compliance with such criteria is a difficult task indeed, in particular if radioactive products from the explosion are not available. This brings up another, even more serious, point: The Limited Nuclear Test Ban Treaty of 1963 forbids explosions which carry measurable radioactive products beyond national boundaries. Now the problem of policing the new Moscow Treaty seems to require that the old 1963 treaty be violated or weakened by amendment! Moreover it has long been recognized that viable nuclear excavation projects could not be carried out within the present terms of the 1963 Treaty. In this context the 1974 Moscow Treaty is thus a real step backwards in arms control. One often mentioned mechanism for policing peaceful nuclear explosions would be a requirement to make peaceful explosions "too awkward" to make them suitable for military use. Again I cannot go into detail here as to what this means beyond expressing a great deal of skepticism.

The greatest objection to the peaceful use clause in the Moscow Treaty is the cynicism which this exclusion presents to the non-nuclear world about U.S./Soviet intentions to really work towards cessation of nuclear tests. The world now sees the very spokesman who objected to a comprehensive nuclear test ban by inventing very elaborate means of conjectured Soviet cheating apparently willing to accept the much larger loophole implied by permitting "peaceful uses". The recent Indian explosion has reinforced that cynicism, since the official Indian position remains that their nuclear devices are solely "peaceful"; one is reminded of Andrei Vishinsky's announcement in 1949 to the United Nations that the Soviets would dedicate their nuclear explosives "to move mountains and to dam rivers".

The very fact that apparently the Soviets and Americans found it necessary in the 1974 Treaty to make an exception for peaceful uses of nuclear explosions indicates to non-nuclear nations that such uses must have enormous significance to the two powers, a conclusion totally unsupported by technical fact. Therefore non-nuclear nations would feel both economic and prestige pressures to join the nuclear club such as India has now done. **In summary. I consider the Moscow Threshold Test Ban Treaty a major disservice to the course of stemming nuclear non-proliferation — it is this problem which is possibly the largest nightmare facing this world today.**

I hope that these two very recent examples of negotiated arms control have illustrated that the results can be either good or bad in the course of halting the ever-increasing danger of nuclear destruction. It is also clear from these examples that whatever good comes out of negotiated arms control agreements can be — but does not have to be — negated by the concomitant effects which have resulted from domestic pressures accompanying arms control negotiations. **I believe that only restraint by all parties on the broad front of nuclear armament, combined with steps in negotiated arms control, can reverse the potentially disastrous course of the last decade.**

11

The Third Culture

David Hawkins

University of Colorado, Boulder, Colorado

1. Introduction

My theme in the Symposium will be evolution. I want to talk about evolution of the grand scale, indeed the very grand scale. That part is just short of metaphysics. I also want to talk about the process on a very small and local scale, which happens to be our own. That part, I hope, is ethics.

This *Festsprach* for Victor Weisskopf is a very suitable occasion, for me at least, to discuss this topic. Physics may be the last of the unhistorical sciences; it is at any rate powerfully relevant to them all. And I can introduce my subject on both levels, the greater and the lesser, by telling a story. It refers to a book, mentioned yesterday by Dr. Killian, classed as popular, written by V. F. Weisskopf, called *Knowledge and Wonder*. Popularization is a bad word because it usually treats science badly in the process, and is treated badly by most scientists. What the word originally meant and literally ought to mean is something you do to science in the interest of people, not what you do to people in the interest of science. They have to do that themselves with what you provide. To popularize science today requires real depth of understanding both of science itself and also of the complexity and subtlety of non-scientific minds, children's or adult's. Instead of talking about "popularizing" science I shall talk about reconstructing it for maximum accessibility. That is what good scientists do and ought to do anyway. They are after all only somewhat deviant samples of the species, and what makes them intelligible to the innocent is of necessity a matter of self-enlightenment as well.

You shall see I am here trying to popularize the idea of a certain kind of scientific popularization. I shall however have to go in a big circle to that. Yesterday in Edward Purcell's lecture I found not only a beautiful example of how to make science genuinely *popular*, but also I found a completely new description to fit my own trade of philosophy: It is a form of life at a very low Reynolds number, and I hadn't realized before philosophy's affinity with the *Escherichia coli*. More seriously I recognized immediately two very apt criteria which fit the behavior of philosophers, as well as *Escherichia coli*. One is that to get anywhere in philosophy your logical configuration must go in circles; the other is that even when you do that it's not easy to decide whether you're going forward or backward. Finally I should remind you that *Escherichia coli* is in origin a colon bacillus of mammals, and that is why, in such company as this, philosophizing takes a lot of guts.

So after a big circle I shall come back to the matter of popularization, and probably go backwards in the process. The story I have to tell, in the meantime, concerns a dialogue over a topic in *Knowledge and Wonder*, the quantum ladder. The quantum ladder describes the stability of the particles of matter, a stability first recognized, in Greek science and in the 19th century, by the concept of the atom as a particle which is simple and indivisible and acts as a whole without troubling us with any internal disturbances in the process. Of course the principles of quantum physics allow us to re-make this concept of atomicity into a relative one, so that particles will appear as simple below a certain energy threshold but as compound or at least complex, above that threshold. I believe that the quantum ladder has contributed one of the most important perturbations to philosophy which modern physics has produced, comparable to and in a sense more important than relativity, or the other consequences of the uncertainty principle. It represents a new alternative, an escape between the horns of a dilemma fought over in Greek times by Stoics and Epicureans, and more recently in the names of Newton and Leibniz. Both sides were wrong, and both were right. It should have been a philosopher who thought of it.

When I had read Weisskopf's excellent exposition of the ladder of nature, of the inverse square dependence of quantum-stability on the size of particles, I pointed out to him that his expression, "the ladder of Nature", was not new in intellectual history, but that the *scala naturae* has a central place in the long tradition of neo-Platonism, from which it eventually provided one of the early pre-Darwinian models, of biological order. There are, to be sure, some important differences. At first sight the differences are so vast that a pedestrian spirit would see in the coincidence only an accident of words. The ancient ladder is a ladder of perfection, it descends to say from heaven. The quantum ladder grows up from the smallest particles and we lose sight of it for most purposes before we get to everyday things, though it gives us the stable world of matter, which was what made Lucretius argue for atoms. Such coincidence may however be deeper than it seems.

To clinch my point I gave the author of *Knowledge and Wonder* a copy of Arthur Lovejoy's *The Great Chain of Being*, a scholarly history of neo-Platonic thought. A good many months later I asked about the book, and the answer was, "Oh, I didn't read it." And then, alluding to C. P. Snow's famous discussion of the literary culture and the scientific culture, he added: "I am a second-culture man; I don't read books." (He only writes them.)

So that is where I got my title for this talk, The *Third* Culture. C. P. Snow — like James Bronowski and a good number of us here — wanted to officiate at a re-marriage of the two cultures — the humanities and the sciences. But I would hold this to be possible only within the framework of a third culture, still struggling to be born, a majority culture, not a minority one like these two — and in many ways unlike the present dominant culture of science, or of literature. The framework for the third culture, I hold, is a whole system of thought habits and sensitivities which defines the human condition in evolutionary terms. If this sounds like a regression to the 19th century, I have been understood — in part.

2. Evolution

The other point of my story is that the quantum ladder leads, in a mind sufficiently prepared, to a certain general view of the stability not only of crystals but also of living forms. Even the smallest living things exist well above the level of quantum stability, and if they did not they wouldn't be living, they are the antithesis of atoms, more complex internally than in all their external relations. They are unstable, and they die. But they are tokens of a type, and the type itself has an extraordinary stability which depends on two conditions. One is the fact that their number doubles fast enough to compensate for death, the second that it does so without change of type. Seeing the theoretical centrality of this fact of biologically identical self-reproduction was indeed Aristotle's great contribution, and it is a pleasure for a philosopher to hear Max Delbrück bestow the credits properly. In a modem view it is the quantum ladder which explains the chemical bond and which makes possible the stability of the genetic code, satisfying the second condition, and this connection is explicit in *Knowledge and Wonder*, whose author doesn't read Plotonius, he doesn't even read Aristotle. Anyway, that is where the two ladders join. The quantum ladder is a ladder of energy barriers. The ladder of life is a ladder of what Aristotle called *information*, and the quantum ladder makes it possible.

But the whole subject goes farther. Aristotle not only discovered DNA but in the process he also proved the impossibility of evolution. Indeed it is only since the recent *re*discovery of DNA — to use the Delbrück manner of speaking — that some slight holes in Aristotle's anti-evolutionary argument have begun to be properly defined. I think it was part of the pall cast by the Bishop Wilver-forces and the Scopes trials and the rest that so many biologists have failed to learn about or understand the full force of the Aristotelian argument, which they deprecated, and knew only in a confused and vulgarized form, through the 19th and 20th century opponents of evolution. In the first approximation evolution really is impossible. It is only the overwhelming historical evidence, and since Darwin our beginnings of a second-approximation understanding of it, which has saved the day for evolution. But in saving the day we surely confront a new one, and that topic is one around which we have had many discussions, ranging from what has come to be called informational thermodynamics — which in qualitative form was basic to Aristotle's thinking — to the present state of the world when seen in the grand perspective of biological history. I shall not dwell on that several-billion year epic, but let me remind you of its outline. Solar cooking generated self-reproducing structures and clusters. These in turn generated living things, things already complex like the algae which left their traces in the pre-cambrian rocks. Living things generated the embryological pathways toward complex organisms, and these generated brains. The evolution of brains generated a marvelous profusion — along one special track it generated what might be called the embryology of culture, uniquely in the form of a fully social mammal. Cultural evolution, again along one very special track among many other marvels, generated science.

I hope this quick summary suggests something of the framework of the Third Culture. Everything in it is a product of the 19th and 20th century revolution in the historical sciences. To elevate the historical sciences to a position of pre-eminence in reconstructing our culture is to make room for the two old cultures, for the arts and humanities as well as for what most of you would more or less automatically mean, as did C. P. Snow, by the honorific word "science". Physics has been the leading edge, one sometimes emulated well and often badly indeed. It has become, surely, an enormously important tool in all historical investigation. Many of the investigations which have sprung from physics have created new dimensions of history. The extended terrestrial time scale came first from physics and astrophysics, as did the space scale. But in the process astronomy itself has turned historian.

Yet I propose to demote physics in the sociological pecking order to a status lower, or at least no higher, than the historical sciences. I take courage, here, from the thinking of Weisskopf and others of you, especially Delbrück (the leader who first went over the fence) and Morrison (who first taught me how, thermodynamically speaking, history is made). Weisskopf is incurably a physicist, I think, but he lives with history, more "funeous" than most and more determined to elevate man's historical concerns among his fellows.

3. The Evolution of Knowledge

In evolution there are certain kinds of order which might be called thematic regularities or analogies. All things new begin as fluctuations within some larger population, and are at first as easily wiped out as begun. In biology they appear first as varieties within some gene pool; in the history of culture they are varieties within a society's wider pool or fund of material culture and ideas. Their development within the pool is often slow and unnoticed, but on occasion they prove preadapted to some favorable set of available circumstances which, in retrospect, evolutionists call an ecological niche. There they flourish and become conspicuous, in some isolation from the maternal pool; they generate new varieties, and the process continues. It is so in all the beginnings of science. Knowledge is first generated and passed on as an offshoot and accompaniment of the arts. This is not to say that its origins are solely utilitarian; we should keep in mind the double meaning of "art". What Cyril Smith has said about the history of materials is true also of the history of ideas; in their beginnings we should probably find as motivation something more like curiosity, speculation (which means reflection, mirroring), an interest and a satisfaction close to the esthetic. But ideas have to reproduce themselves in the minds of successive generations, and the niche for them to become stable and continue is not available except as they find place and utility within some ongoing activity or vocation. At that stage science *is* still popular. The social history of science leads us back, usually, to the relatively late times and circumstances when such ideas penetrate, prove pre-adapted, to the minority culture of scholars, lectures, students and books, where transactions and accommodations occur between the new subject matter and the traditions of learning. We find this sort of penetration and fusion in all the beginnings of serious self-conscious investigation, taking off from carpentry and masonry, from navigation, from mining and metallurgy, from weaving and pottery, from soothsaying and dice gaming, from the market place, — all of which had a long submerged life invisible to the learned world, before the knowledge evolved in them came to be taken up and certified and extended by the professional knowers. Indeed, who can say, today, what beginnings of new knowledge may be implicit, mute and unnoticed, in the vast range of twentieth century experience?

After many such incorporations, and through the symbioses and fusions of ideas, science has emerged in its distinctive modern form as the institutionalized deliberate use of knowledge to extend knowledge, transforming what once had been a by-product into a central aim, and transforming practical application itself (if you can get away with it) into a subordinate means. As part of this evolution knowledge has developed certain characteristic styles of organization and codification which for most of us here are close to our daily bread, and which are so pervasive that we sometimes make the error of supposing that they are intrinsic to subject-matter itself.

Such styles of organization have indeed evolved, along with scientific inquiry, in the interest of clarity, coherence and applicability. They vary, of course, from stage to stage and subject to subject, but have a strong family resemblance. That organization has also another function inseparable from the first, to provide more or less standardized pathways for teaching and learning. These two aims of organization, for application and research on the one hand, for teaching and learning on the other, are rather different in the demands they make, but these conflicting demands are partially reconciled in a rich context which incorporates much else besides the formal organization implied by the textbooks and lectures, the monographs and journals. It has to be a medium, not a vacuum, and one with a complex structure. It is a fully human context of informal association in the apprentice-ship and collaboration of masters and students. As we all know the organization of knowledge codified in scientific literature, if *put* in a vacuum, is an impoverished and in many ways radically misleading guide to the intellectual structure of a good scientific mind. In the world of literature books are the final product of a high art. In science they are rarely that; more essentially they are tools of the trade, hardly ever read from cover to cover, valued rather for what the computer trade perversely calls *random* access. That is why it can *be* a joke to say that persons of the second culture do not read. Textbooks may in their own way be works of art, but the genre is not like that of literature or history. The textbooks is more like a well-stocked and well-organized shop, written with the clear anticipation that there will be a competent master around to induct the apprentices into its uses. You can in fact *read* a textbook only when you already know most of its content. That is true also, some of us must confess, of symposium lectures.

Because of the rapidity of scientific evolution in the niches which modern society has provided it, the written and spoken communication of scientists, considered from the point of view of the outside world, has become profoundly inaccessible and esoteric. It's organization makes sense only within a rather complex sub-culture which occupies much of the waking lives of its participants, pre-supposing years of steady involvement and discipline as a pre-condition. In flourishing, science constantly pulls a small subset of persons *into* itself *as* a minority culture, not, in fact, through a very wide spread of that enlightenment to which, historically and morally, it is in principle so deeply committed.

4. Unevenness — Esotericism

I think it is right to say that this failure, of one of the golden dreams of the enlightenment, is real enough to require acknowledgement even after all the necessary qualifications are in, and in spite of all efforts from within science both to reflect upon its own rapidly evolving treasure and style, and to make this understanding available to a wider audience within other minority sub-cultures and to the world at large. The esoteric isolation of scientific knowledge and style has not been a conspiracy of self-protection, but it is there and growing nevertheless, a quasi-conspiracy. It is hard to name any of the intellectual heroes of modern science who have not at some level made major efforts against this tendency, from Galileo to Einstein and Bohr. Most of the great figures in modern science have in fact also figured greatly in the history of modern philosophy. In asserting the present-day cultural isolation of science I think I can disarm any charge that I am joining the opposition. I was myself imprinted and trained up within this traditional enterprise, and my own heroes remain those of the scientific enlightenment. To be even more disarming, I should mention that philosophers can be even more esoteric than Julian Schwinger. All noble things, said Spinoza, are as difficult as they are rare — to start with.

There is of course one kind of channel of communication which the rapidly rising level of scientific knowledge has not constricted but only deepened, its interaction with industry, with warfare, with agriculture and medicine. Some at least of the promises of Francis Bacon have been delivered in full measure, pressed down and running over, and the world will never be the same again. Following another one of those thematic analogies from evolution, the rapid growth of the new cultural variety called science has transformed the very character of the system within which it first found a place to flourish. In *that* direction the structure of knowledge and the direction of scientific endeavor has not proved unintelligible or esoteric. So the larger subculture of technology and industrial craftsmanship has evolved with science, has contributed to its material instrumentation and also to its new *conceptual* tools. M.I.T. can well boast, and then worry, about that contribution. This instrumental aspect of science has been so much at least of its essence that its opponents, from the traditions of religion and philosophical idealism and the older humanities, have plausibly and gleefully condemned it to be *only* instrumental or operational in character, without status as the kind of knowledge to throw light on the nature of things, on man's place and potential, or to inform the ends of our existence. And not infrequently the spirit of academic positivism has welcomed that opinion and has constructed a system of judgments according to which science is in fact harmlessly value-free and therefore *ought* to be left alone just to do its own thing. But such judgments are themselves value-loaded.

Considered abstractly, knowledge about nature and its workings is *per se* value free in an obvious sense. There can be no more doubt about this than that number has no mass, that the Euclidean triangle is neither red nor green, or that love cannot be conveyed by a pill. But this philosophical profundity cannot be converted except illicitly to a lack of value-relevance.

For while facts in this abstract sense do not *entail* values, values on the other hand do entail beliefs about matters of fact at all levels we care to explore. When therefore new knowledge casts doubt upon settled factual belief there can be hell to pay and sometimes, also, the intimation of new heavens, previously inaccessible, to be gained. I remember a discussion in Los Alamos, immediately postwar. We were discussing the resistance of many people to disturb their own settled priorities about the propriety and inevitability of war, thus declaring the moral and political irrelevance of those new famous facts we were concerned about. It was Victor Weisskopf who produced a terse summary about the difference between sufficient conditions and necessary conditions in that context. I quote from memory:

"It is true that facts are not enough; but sometimes they are too much!"

If the evolution of knowledge within science has deepened the channels of its communication on some instrumental levels, it has tended to choke them on other levels. Historically I suppose this is unavoidable, it is a direct result of differential rates of evolution. Yet it produces something which can act as a poison within the whole of society, being, at the same time, of the greatest potential benefit. It is much as the early green plants may once have poisoned the atmosphere with what is for us the blessing of oxygen. It produces a society which is irreversibly committed to the fruits of science and technology, while increasingly alienated from them on an intellectual and moral and esthetic level. This alienation, the expression of distaste and misgiving at the very least, has been around for a long time, but it has been restricted to other minority cultures until recently. You can find it in John Donne before Newton, in the great Pascal, in theology and philosophy all along, in Victorian poets and in some powerful writers of our century. A critic who worked for a time in M.I.T.'s humanities program has documented much of this in a book of essays taking off the Snow–Leavis controversy: Martin Greene, *Science and the Shabby Curate of Poetry*. I would only remind you that those anti-scientific attitudes which were elaborated within a high literary tradition which was itself esoteric — that of Pound, Eliot, and Lawrence — have not become commonplaces, a dime a dozen, along with much else we try to listen to, of the college generation (I suppose even M.I.T. undergraduates are not immune).

But that is not the main point. I think what we must realize is, mainly, that science is being seen only from the outside through its expressions in industrial and military technology, but — may I say it — reinforced by a personal reading of the moral character of science as not only operational in origin but as arrogantly unconcerned or arrogantly manipulative, as linked by both nature and ambition to those centers of manipulative power — what has come to be called "the establishment" — which our society produces with some abundance.

I do not want to let physics and biology off this hook but I think the main source of that popular reading of science is now also becoming a dime a dozen. It is the narrow view, propagated by some philosophers and taken up as methodology in some fields, which I have called academic positivism: science is value-free, leave us alone. Our knowledge and skill will be available to all bidders, whatever their values, as a technique of manipulation but not as informing their goals. So the view that scientific knowledge is only instrumental, and that it has nothing to do with the definition of ends, gets reinforced by some of those who speak in the very name of science.

5. A New Direction for Science

But all of this, it seems to me, is only a surface manifestation of the kind of evolutionary pattern I have wished to remind you of. I repeat. Because of the rapidity of its evolution and the way it has been geared to that evolution as a deliberate aim, science has cut itself off increasingly from pervasive influence within the common culture; it has not done so as a matter of intent, but, in a transient historical sense, of necessity. Indeed it has done so contrary to its own moral traditions of enlightenment and the efforts of many to extend those traditions; and this fact becomes increasingly a stumbling-block in the very career of science. Forty years ago John Dewey, himself one of the most persistent and penetrating advocates of scientific enlightenment, spoke of the fact that the impact of scientific knowledge on our older traditions has been mainly that of a disruptive agency, seen only through its external manifestations. Not possessing the power of scientific knowledge and procedure, not having truly assimilated those power, such venerable traditions were powerless to redeem themselves against it. Science is yet too new, Dewey said, to have penetrated into the subsoil of mind.

I cannot find a stronger expression than that, to underline my own thesis. But today the isolation which has favored the evolution of knowledge will no longer do so, and this must alter many of the priorities we have grown to accept. This is why I spoke disparagingly, at the beginning, of the usual concept of popularizing, which typically implies a kind of expository writing or lecturing which somehow *translates* scientific thinking, or investigation, or conclusions, into non-technical prose assumed accessible to someone called a layman. Very simply, it does not reach anywhere near what Dewey called the subsoil of mind. No mere discourse will in fact do so, except in a context which *as* context invites learners to an active process of reconstructing their own intellectual apparatus, frequently at a level which would be judged, by typical scholastic standards, to be almost unattainably elementary — elementary in a respectful sense, that of the elements which discourse itself, in science, presupposes. The task of penetrating to this level is one which requires talents prized in the scientific community and therefore difficult, under the present professional ethic, to come by. Under these priorities the kind of research and practical investment I have in mind will all too often be regarded by the ambitious young ones as low level in character, fit for professors emeritus (I hesitate to use that phrase), at best as an amiable eccentricity on the part of those who engage in it. But this work also requires a serious amount of experimental participation in teaching situations, to further it. It could be called popularization if — to repeat — that term were taken to imply a very radical reconstruction of the organization of scientific knowledge, with the aim of maximizing its penetrability from the outside and its assimilability either by minds whose powers are first developing, or which have developed in other patterns than those now deemed apt for science. I could use a fashionable term and speak of maximizing the interfacial area between scientific minds and those others, with the aim that each could better reconstruct itself. I could use the image which Jonathan Swift conjured up as a literary man's vision of the kingdom of heaven, where persons are said to be books open to one another. I could use the imagery of the biological molecules which Philip Morrison has popularized for me, of one-dimensional helical structures which are never buried irretrievably in any three-dimensional matrix, but which can also, as occasion requires, be open to one another.

I propose to illustrate the value, for this general aim, of searching for and defining those almost irretrievably elementary stumbling blocks which pedagogy normally sweeps under the rug because it does not understand them, and by which as we all at time know it is normally embarrassed and disgruntled.

I shall try to suggest a few random examples of what I mean. These have come from a pretty varied experience. I haven't kept statistics, but the conceptual and perceptual troubles they reveal will show up in almost every group of students and adult teachers and in most children as well, especially if they're lively enough to reveal them to you, or you are smart enough to find them. The first example is from biology, and it seems unaffected by the presence or absence of one or two high school or college courses. Trees grow *up* out of the ground, raising their branches with them, and the new wood each year is either diffused through them as they somehow swell, or is inside, in the heartwood. I had two college students, future teachers of the young, working on this for many hours and to my amazement all the evidence they could find, from dissection and observation, only confirmed their settled view. I have repeated this diagnostic test many times, and the result holds up. I can imagine no greater insult to the style and dignity of a green plant than thus to confuse its growth and form with that of animals. But I ask you to remember how recent is the knowledge that plants are made of air and water and sunlight, and how much acquaintance with the great discoveries of the 18th and 19th centuries is required to feel comfortable with this allegedly simple fact. If it is not simple, we have only forgotten *why* it is not. That context has not penetrated the subsoil of the mind.

Part of that context is much older, it depends on some free and full credence given to the material reality of air, my second example. Twelve years ago in Watertown some of us naive ones began to learn with delight about the unreality of air. Given an empty test tube inverted in water, and a syringe with a little plastic hose pipe on the end of it, you are asked to fill it with water. A majority, of all ages, will try to fill it by filling the syringe with water, snaking the tubing into the test tube, and pumping water in. Why does not the water rise? The tube is, after all, *empty!* There are many other ways of confirming this non-entity of air, and one came up recently in a group of teachers: They were about to put some alka-seltzer tablets in the water under the open end of a very long plastic tube, corked at the top and this time almost but not quite full of water. A discussion arose among our teachers — some of whom dearly *want* to understand things — as to whether the bubbles ascending in the tube would *raise* or *lower* the water level in it. A unanimous consensus was that the water level in the tube would be *raised*, not lowered. Later discussion revealed the logic of this inference, one which mystified me and then reminded me of *my* persistent ignorance. I who played with chemical apparatus at the age of eight. The bubbles rising take up space so they push *up* on the water. Not bad, so far. When they reach the top the bubbles break and — cease to exist. One is reminded again of the recency of scientific thought and of its encapsulation within a narrow tradition — in this case the recency of Plato' and Galileo's and Toricelli's ocean of air. Another question which came up I leave with you. Can you make a siphon of sand?

Third example: the sun changes place during the day, but it is not *moving*. Fourth example: when you see things in a mirror, you see them where they see themselves, on the surface of the mirror like a picture. This one again, once you start to unpack it — as every physicist knows — is a marvel of the mechanics of waves. But even in its single geometry it is a marvel of context. It involves the odd notion that mirrors are wrong-way windows. The picture hypothesis or the mirror-world-isn't-real hypotheses is 60%–80% correct in predicting responses of graduate students in departments other than physics and mathematics. There are some further notions about vision which, called "scientific", the eye does not tell itself and even seems to contradict. Generations of philosophers, until recently, until the coming of what might be called the thermodynamics of vision, have fought a losing battle against the proposition of "popularized" optics, that because the eye is a camera the world we see is somehow an affair of images inside the head. In this case as in many others common sense, suitably analyzed, could long since have given the lead to science — except that until recently the opportunism of scientific evolution could not yet afford it a hearing, and got bowdlerized from a system of true propositions about some of the *means* of vision to a system of false propositions about vision itself. And there are many other battles, yet to be staged, perhaps an infinite number, between those two kinds, of knowledge which lie on opposite sides of our ignorance — organized empirical science as it evolves on one side, and on the other that more mysterious sort of knowledge which we already have but tend to know about only in fragments hard to analyze — the kind for which Plato and Aristotle, in our own tradition, invented the mirror of philosophy. The reconstruction we must be committed to is not along a one way street and is sometimes deep.

Let me return to a second round of my circle, the grander scale of evolution. From the carbon molecules to the cells, from embryology to the brain, from self-reproducing culture to expanding science, each level of evolution has altered the conditions of those which preceded it, *and* altered their direction. In modern times an increasingly dense or frequent series of developments represents the latest but one of the evolutions we most consider, the way science has altered the boundaries of cultural evolution generally. From now on we can plot another series, however, that of the *direct* influence of physics and biology on the terrestrial gene pool and *its* evolution. As Robert Oppenheimer once said in connection with nuclear weaponeering, decisions grow pretty inevitably out of possibilities. Perhaps it was a self-justifying remark, it was not a false one. In any case I move it over to the realm of the new biology. We talk about growing systems of neutrons *in vitro*, and Arthur Clarke has already given that technology a name; the results are called biots, biological automata. They may not be willing to be what we now call computers; more likely the style will be more interesting than that of our present hardware. For another example, there is a small cloud on the horizon when people begin to worry that a new kind of flu epidemic may have already once escaped from the laboratory. A world which doesn't yet understand the growth style of trees is not equipped, morally or in any other way, to understand the ethics of such concerns. Persons within science are better able, as are persons who have genuinely assimilated its culture, even at relatively elementary levels. But unless scientific knowledge and ways of thought can be made accessible and absorbing to soon-growing circles within the population, the chances are all too great that the engineering of the biological nucleus will join that of the atomic nucleus, and still older forms, in unknowingly bringing about major and irreversible perturbations to the biosphere and then, inevitably, to the neosphere as well. I respect Hans Bethe's cool judgements about energy problems and very much his partisanship for heavy water reactors against others available. And I am very much a skeptic about those solidified radioactive wastes, and the meltdown problems. I think they both have to be considered on another time scale, where both materials and human institutions may look somewhat different from the way they appear on present-day 50 year planning horizons. But that needs a symposium of its own.

For the atomic nucleus, at any rate, we need no longer prognosticate, the perturbations are already upon us, and, after all, our energy problems are dwarfed by the obscenity of the world's nuclear stockpile, which Hans Bethe and others here have worked so hard to help make sense of. For the biological nucleus, apart from all the indirect consequences of population growth and industrial growth, we already have some to worry about as direct consequences. Bacterial selection against antibiotics is probably worth the price, and the price finite; the ecological consequences of the engineering of high yield grasses with very limited genetic variety can perhaps be remedied fully.

Yet it seems to me there is a kind of theorem — perhaps only another thematic regularity or analogy — which modern biology suggests. Living things — cells — are highly self-organizing systems. By their very nature they will seize and exploit any fresh opportunities we provide them, bringing to the task an ancient and very complex molecular wisdom which we can abstractly understand but which we surely cannot replace *de novo*. It is too complex and the time scale of its acquisition is too long.

But we can, *much* more easily learn to channel it. In this domain surely, as all too probably with most long range human decisions, what is possible becomes actual without adequate ways of looking for consequences. So our intervention in the cell's nucleus should probably not even be called engineering at all, but rather a new kind or level of evolution, intended or not, one speeded up by a good many factors of 10, a time scale for biological change accelerated to that of culture at its speediest. Up till this time the human effect, radical though it is, has been mainly to destroy old and create new ecological loci. The cells have managed so far to get along with their slower kind of learning, that of variation and selection. When however we intervene directly with *our* own kinds of well-tailored macro-mutations, reproducing themselves *ad lib*, the whole story of evolution begins not only a new essay, or a new volume of essays, but a whole new series with the subtitle, *post-Darwinian.*

With that I come back to the local time and scene. The great danger in our present world is that we will be clever in detail but will not see or direct our cleverness within an intellectual and moral framework ample enough to even imagine the role which evolution has forced upon us, that of responsibility for a very complicated and some what messy planet, *and* from here on out.

As the institutions of science and its knowledge-structures are now constituted, we are relatively powerless to determine even what we, ourselves, in science, will do. We confront a spectrum between two alternatives.

One is that of a small, therefore in the end corruptible and in any case incompetent technocracy. The other is that of a new and major effort toward the reconstruction I have spoken of. I picked a few examples to suggest one level — that of the almost unattainably elementary — on which such reconstruction must go to work. At a level slightly higher, the context for any intelligent and persistent *evolutionary* thinking is almost lacking in our society. A sense for the tempi and modes — in the phrase of G. G. Simpson — of evolution is mostly not accessible in the school or the market places, and even most biologists, counted over the range of a sizable profession, are not prepared to take themselves or their students very deeply into Darwin or much beyond him. Good that the band which *are* prepared is now again growing and with the tools of modern biology. There is need for reconstruction at all levels.

Much as I confess I like the image, not all scientists should take a year's apprenticeship as teachers of young children or of *their* teachers. We need to protect, as well as teach! But many *should*, I think, and many more would at least profit themselves if they did. But still others are needed to help those more directly involved; a new ladder from heaven. To illustrate *that* kind of help, and to extricate myself from the rather sober style in which my thinking seems to have entrapped me, I will tell a final story. A good many years ago in Boulder, a seven year old asked me, "Daddy, why are the mountains *that* high (pointing)?" I was at a loss, as parents and teachers often are, not so much for sure answers as for the context of such a question, and for ways of giving *it* adult dignity and value, rather than squelching or evading. A little time later I passed this question on to Victor Weisskopf. I had thought about it in terms of its baptism by the geologists of that time, including the measured deformation of rock under high pressure, isostasy and the rest. Complicated, but Viki chose the easy way, without even pulling from his wallet that card of dimensional constants. Soon we had gamma, the mass of the earth, and the magnitude of the van der Waal's forces — an answer came out, as I remember, which not only fitted our mountains within a factor less than 2, but also did well for the moon and for finding the biggest asteroid shaped like a can of soup.

I can't directly transmit any of this to my friends, the teachers; the context is not theirs yet. Like many other good examples of simple science, it still needs pretty radical reconstruction in the circle of persons I have chosen for my stories. The ladder reaches down from heaven, but not far enough. It simply gives me a sense of style and knowledge to use to dig deeper, to try to reach those almost unattainable elements, and to find in teaching and their children, the strengths of mind we here have mostly forgotten we once used in building them. Around the world, some in and some not too far from M.I.T., there is a hardy band of, persons trying to renew and make more visible the romance of education in science, with much detail and some, at least, of worthy generalization, busy working at the reconstruction which will get science deeper into the subsoil, the cultural inheritance of the race. For that is where the kind of courage can be generated which Plato defined as the *knowledge* of the grounds of hope and fear.